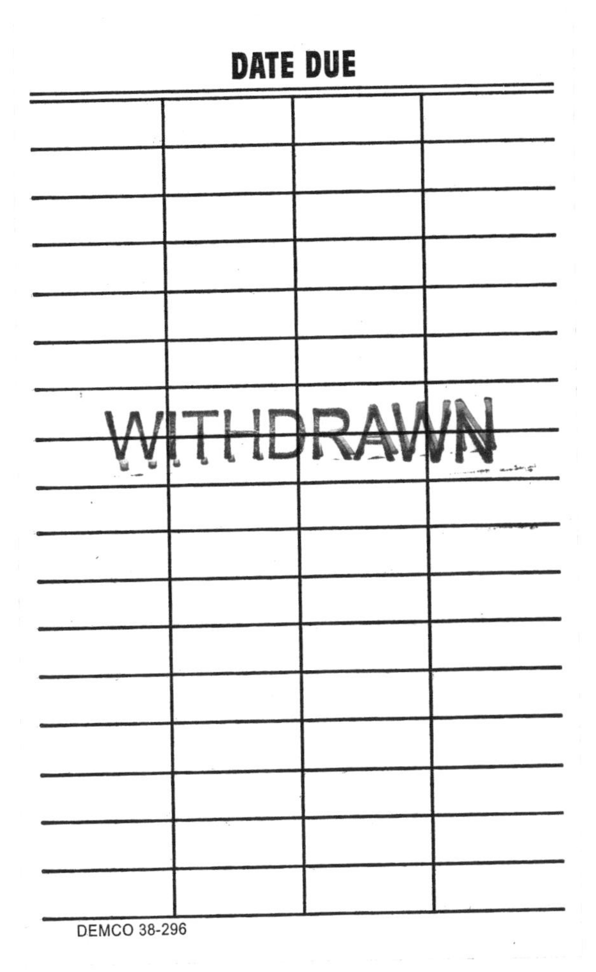
DATE DUE
WITHDRAWN
DEMCO 38-296

FASKEN LEARNING RESOURCE CENTER
9000087618

XCOR, Developing the Next Generation Spaceplane

Erik Seedhouse

XCOR, Developing the Next Generation Spaceplane

Published in association with
Praxis Publishing
Chichester, UK

Erik Seedhouse
Assistant Professor, Commercial Space Operations
Embry-Riddle Aeronautical University
Daytona Beach, Florida, USA

SPRINGER-PRAXIS BOOKS IN SPACE EXPLORATION

Springer Praxis Books
ISBN 978-3-319-26110-2 ISBN 978-3-319-26112-6 (eBook)
DOI 10.1007/978-3-319-26112-6

Library of Congress Control Number: 2016930303

Springer Cham Heidelberg New York Dordrecht London

Cover designer: Jim Wilkie
Project copy editor: Christine Cressy

Printed on acid-free paper

Praxis is a brand of Springer
Springer International Publishing AG Switzerland is part of Springer Science+Business Media (www.springer.com)

Contents

To: Jeff Greason and his team of dedicated engineers for bringing the Lynx to reality

Acknowledgments

In writing this book, the author has been fortunate to have had five reviewers who made such positive comments concerning the content of this publication. He is also grateful to Maury Solomon at Springer and to Clive Horwood and his team at Praxis for guiding this book through the publication process. The author also gratefully acknowledges all those who gave permission to use many of the images in this book, especially XCOR. Thanks also to Jeff Greason for agreeing to be interviewed for this book and to Steve Heck, Jason Reimuller, and Rick Searfoss, who provided valuable insight and input.

The author also expresses his deep appreciation to Project Manager, Sasi Reka, to Christine Cressy, whose attention to detail and patience greatly facilitated the publication of this book, and to Jim Wilkie for creating yet another striking cover. Thanks Jim! Thanks also to Vasco for allowing me to include the material in Appendix IV and to Justin for allowing me to include the material in Appendix V.

About the Author

Erik Seedhouse is a fully trained commercial suborbital astronaut. After completing his first degree, he joined the 2nd Battalion the Parachute Regiment. During his time in the "Para's," Erik spent six months in Belize, where he was trained in the art of jungle warfare. Later, he spent several months learning the intricacies of desert warfare in Cyprus. He made more than 30 jumps from a Hercules C130 aircraft, performed more than 200 helicopter abseils, and fired more light anti-tank weapons than he cares to remember!

Upon returning to academia, the author embarked upon a master's degree which he supported by winning prize money in 100 km running races. After placing third in the World 100 km Championships in 1992, Erik turned to ultra-distance triathlon, winning the World Endurance Triathlon Championships in 1995 and 1996. For good measure, he won the World Double Ironman Championships in 1995 and the infamous Decatriathlon – an event requiring competitors to swim 38 kilometers, cycle 1,800 kilometers, and run 422 kilometers. Non-stop!

In 1996, Erik pursued his Ph.D. at the German Space Agency's Institute for Space Medicine. While studying, he found time to win Ultraman Hawai'i and the European Ultraman Championships as well as completing Race Across America. Due to his success as the world's leading ultra-distance triathlete, Erik was featured in dozens of magazine and television interviews. In 1997, *GQ* magazine nominated him as the "Fittest Man in the World."

In 1999, Erik took a research job at Simon Fraser University. In 2005, he worked as an astronaut training consultant for Bigelow Aerospace and wrote *Tourists in Space*, a training manual for spaceflight participants. Between 2008 and 2013, he served as director of Canada's manned centrifuge and hypobaric operations and, in 2009, he was one of the final 30 candidates in the Canadian Space Agency's Astronaut Recruitment Campaign. Erik has a dream job as a professor in Commercial Space Operations at Embry-Riddle Aeronautical University in Daytona Beach, Florida. In his spare time, he works as an astronaut instructor for Project PoSSUM, a professional speaker, a triathlon coach, and an author. *XCOR* is his 24th book. When not enjoying the sun and rocket launches on Florida's Space Coast, he divides his time between his second home in Sandefjord, Norway, and Waikoloa on the Big Island of Hawai'i.

Acronyms

AASA	Axe Apollo Space Academy
AEM	Animal Enclosure Module
AFT	Autogenic Feedback Training
AGSM	Anti-G Straining Maneuver
AIM	Aeronomy of Ice in the Mesosphere
ARC	Ames Research Center
ATV	Atmospheric Test Vehicle
BPPV	Benign Paradoxical Positional Vertigo
CAA	Civil Aviation Administration
CCL	Commerce Control List
CEF	Change Evaluation Form
COMSTAC	Commercial Space Transportation Advisory Committee
CP	Cowling Port
CRM	Crew Resource Management
CRuSR	Commercial Reusable Suborbital Research
CS	Cowling Starboard
CSF	Commercial Spaceflight Federation
CSLA	Commercial Space Launch Amendments
DARPA	Defense Advanced Research Projects Agency
ECG	Electrocardiogram
ECLSS	Environmentally Controlled Life-Support System
EPT	Effective Performance Time
ERAU	Embry-Riddle Aeronautical University
FAA	Federal Aviation Administration
FAI	Fédération Aéronautique Internationale
FAR	Federal Aviation Regulations
FFD	Final Frontier Design
FOP	Flight Opportunities Program
FRR	Flight Readiness Review

G-LOC	Gravity-Induced Loss of Consciousness
GOR	Gradual Onset Rate
HAI	High-Altitude Indoctrination
HEPA	High-Efficiency Particulate Air
HFA	Hardware Feasibility Assessment
HMD	Head-Mounted Display
HSG	High Sustained G
HTPB	Hydroxyl-Terminated Polybutadiene
ICB	Informal Consent Briefing
ICD	Interface Control Document
ICP	Intracranial Pressure
IPP	Innovative Partnership Program
ISS	Integrated Spaceflight Service
ITAR	International Trade on Arms Regulations
LEO	Low Earth Orbit
LoV	Loss of Vision
LPMR	Layered Phenomena in the Mesopause Region
MASS	Mesospheric Aerosol Sampling Spectrometer
MCAT	Mesospheric Clear Air Turbulence
MCC	Mission Control Center
MCP	Mechanical Counter Pressure
MRI	Magnetic Resonance Imaging
NAUI	National Association of Underwater Instructors
NITE	Noctilucent cloud Imagery and Tomography Experiment
NSRC	Next Generation Suborbital Researchers Conference
PAR	Payload Anomaly Report
PGSC	Payload and General Support Computer
PI	Principal Investigator
PIM	Payload Integration Manager
PLL	Peripheral Light Loss
PMC	Polar Mesospheric Clouds
PMR	Post-Mission Report
PoSSUM	Polar Suborbital Science in the Upper Mesosphere
PSD	Physiological Support Division
PSI	Planetary Space Institute
PUG	Payload Users Guide
RCS	Reaction Control System
RD	Rapid Decompression
REM	Research Education Mission
RLV	Reusable Launch Vehicle
ROR	Rapid Onset Rate
ROSES	Research Opportunities in Space and Earth Sciences
RRL	Rocket Racing League
SARG	Suborbital Applications Researchers Group
SD	Slow Decompression

SMS	Space Motion Sickness
sRLV	Suborbital Reusable Launch Vehicle
SSI	Space Science Institute
SSME	Space Shuttle Main Engine
SSTO	Single Stage to Orbit
SSTP	Suborbital Scientist Training Program
STEM	Science Technology Engineering and Mathematics
STMD	Space Technology Mission Directorate
SwRI	Southwest Research Institute
USAF	United States Air Force
USML	United States Munitions List
USRA	Universities Space Research Association
VFR	Visual Flight Regulations
WFI	Wide Field Imager

Preface

For years after SpaceShipOne won the X-Prize, all you ever heard in the commercial spaceflight business was when SpaceShipTwo would begin revenue flights. Initially, Paris Hilton and her celebrity friends were due to take their suborbital joyride in 2007, but an explosion that killed three workers put paid to that deadline. Then 2010 was announced as the start of revenue operations but, by the end of 2010, still no passengers had flown. 2010 became 2011, which became 2012 and still there were no flights. Then, tragically, in October 2014, SpaceShipTwo crashed, killing one of the pilots and injuring the other. The public wondered whether passengers would ever fly in space, oblivious to the work of a company that also had suborbital aspirations and which was located just a stone's throw down the flight line from Virgin Galactic. That company's name is XCOR and its snappy little spaceship is the Lynx.

The Lynx has been in the works for years, but XCOR, unlike some companies, prefer to let their deeds to the talking. No bold pronouncements of when revenue flights will start from this company. Over the years, XCOR has amassed invaluable expertise in the building of suborbital vehicles: in addition to having developed and built 13 different rocket engines, XCOR has also accumulated more than 4,000 engine firings and more than eight hours of run time on their engines. With the travails of Virgin Galactic putting the future of SpaceShipTwo on a back foot, XCOR has been thrust into the spotlight of the commercial space industry and is on the cusp of conducting flight testing of the Lynx Mark I.

The Lynx has two seats – one for a pilot and one for a spaceflight participant. Its low weight and high-octane fuel confer important advantages over SpaceShipTwo that include direct runway launches without the complication and expense of a mother ship and the ability to fly several times per day. Like SpaceShipTwo, the Lynx is a rocket-powered airplane, but that's about the only similarity. Powered by four XCOR-built kerosene and liquid-oxygen engines, the Lynx's take-off speed is 190 knots, and it can get airborne with only 350 meters of runway. The all-liquid design is more efficient than SpaceShipTwo's hybrid propulsion, providing more thrust per pound of fuel. All-liquid fuel should also give the Lynx a fast turnaround between flights because crews can just top up the tanks and fly again, whereas SpaceShipTwo's engine must be replaced between flights.

Passengers paying US$150,000 ($100,00 less than Virgin Galactic's ticket price) will ride beside the pilot. Both pilot and passenger will wear pressure suits as a safety measure in case cabin pressure is lost during the flight. Unlike SpaceShipTwo customers, Lynx passengers will not be able to unstrap and float about the cabin after the engine cut-off. All being well, revenue flights could start sometime in 2019. That's 15 years after the X-Prize-winning flight of SpaceShipOne and there may be some who are wondering why this suborbital spaceflight business has taken so long. The answer is money. XCOR never had the deep pockets of a Virgin Galactic, a SpaceX, or a Blue Origin. This is a company that has accomplished what many industry wags thought impossible on a budget that NASA uses to put together a few PowerPoint presentations. And it has done so thanks to the incredible dedication and perseverance of a handful of extraordinarily talented individuals who had the intestinal fortitude to take risks and to dream big. Take Jeff Greason for example. We'll talk about Jeff at some length in this book but here's a snapshot of the man with the vision that morphed into what XCOR is today.

The XCOR team. Credit: XCOR

Jeff has been space enthusiast his whole life so, when an opportunity to take the job as head of propulsion with Rotary Rockets came about in 1997, he jumped at the chance. It was a bold – some may say reckless – move, given that he left a lucrative career as an engineer with Intel, but "bold" is what Jeff does. Two years later, Rotary folded and Greason, together with a small group of Rotary engineers, formed XCOR. More than 16 years later, they are still together[1] and are on the cusp of making history as the first company to start a suborbital flight service. And, when that service starts, the pilot at the controls will likely be three-time Shuttle astronaut Rick Searfoss. With Searfoss and his passenger ensconced in their pressure suits, the Lynx will taxi off the ramp and wait for clearance from the tower at Midland. Once clearance has been given, the Lynx will get airborne in seconds thanks to the eye-popping acceleration provided by those engines. Less than a minute after take-off, the Lynx will be accelerating through Mach 1 and the sky that was blue just a few seconds earlier will rapidly fade to black. With the flip of a few switches, Searfoss will shut down the engines and momentum will do the rest as the vehicle coasts to its apex more than 100 kilometers above Earth. There, for up to four minutes, passengers – now astronauts – will take in the jaw-dropping view, unless they happen to be scientists, in which case they will have to knuckle down to following their checklists. All too soon, the suborbital joyride will be over and the Lynx will glide back to its home airport, ready to do it all over again.

[1] In November 2015 it was announced that Jeff, together with two other founders of XCOR Aerospace, were leaving the company to form Agile Aero. While Jeff remains on the board, he is no longer involved in XCOR's day-to-day operations.

1

XCOR: A Brief History

Credit: XCOR

E. Seedhouse, *XCOR, Developing the Next Generation Spaceplane*,
Springer Praxis Books, DOI 10.1007/978-3-319-26112-6_1

In 1999, XCOR comprised four employees who had just been laid off from Rotary Rocket. With no money, no investors, and little in the way of a business plan, they decided to strike out on their own and founded XCOR (*www.xcor.com*). Fifteen years and US$45 million (mostly raised from venture funds) later, XCOR is on the threshold of commercial suborbital passenger operations – all for US$150,000 a ticket.

XCOR has never been a large company, but what it lacks in size it more than makes up for in innovation. While other companies in the New Space era have crashed and burned – think Kistler and Starchaser – XCOR has grown from strength to strength. The reason is simple: XCOR is one of the few companies in the commercial spaceflight arena that can successfully translate their plans to products – a skill they have repeated over and over again since the company's inception in 1999. Back then, the Mojave-based company's project was the NeX-1. The Nex-1, a replica of Chuck Yeager's Bell X-1 (Figure 1.1), was a much more down-to-earth affair compared with the Lynx, since it was merely intended to be shown at air shows. As part of the NeX-1 project, XCOR redesigned the XLR-11 engine that provided the power for the X-1.

Why the NeX-1? At the time, XCOR's plan was to provide high-altitude, Mach-speed joyrides – a precursor to space tourism. It was a bold move back in 1999 because space tourism was a decidedly risky business proposition given that it would be five years before

1.1 The X-1 rocket plane. Credit: NASA

SpaceShipOne made her historic flight. But XCOR's president, Jeff Greason, and a group of like-minded aviation enthusiasts staked their careers on the belief that rocket-powered engines would eventually translate into suborbital joyrides. But, before that metamorphosis could occur, those rocket engines were put to other uses when XCOR introduced the EZ-Rocket to the world. The EZ-Rocket (Figure 1.2), which is discussed in more detail in Chapter 4, made its inaugural test flight on 21 July 2002, flown by Dirk Rutan, brother of the more famous Burt, he of SpaceShipOne fame. Designed to become the catalyst for rocket racing, the EZ-Rocket[1] looks like a far cry from a suborbital spaceship, but XCOR reckoned that developing the technology needed to fly rocket-powered aircraft at insanely fast speeds was a key step to ultimately realizing routine and affordable space travel. And it wasn't as if XCOR was the first to have the idea: before World War II, air racing was an important driver in developing aviation technology. If it worked in the 1930s, then it could work in the 2000s. The EZ-Rocket was basically a revamped Long-EZ aircraft that had had its propeller replaced by a rocket engine which happened to be just a meter away from the cockpit. Before its reincarnation as a rocket-propelled aircraft, the Long-EZ was capable of a top speed of 190 knots but, with its rocket pack, it would have been capable of much faster speeds. But that wasn't XCOR's intention. For one thing, allowing the rocket to operate at full throttle (400 pounds of thrust) would have torn the airframe apart and, for another, the

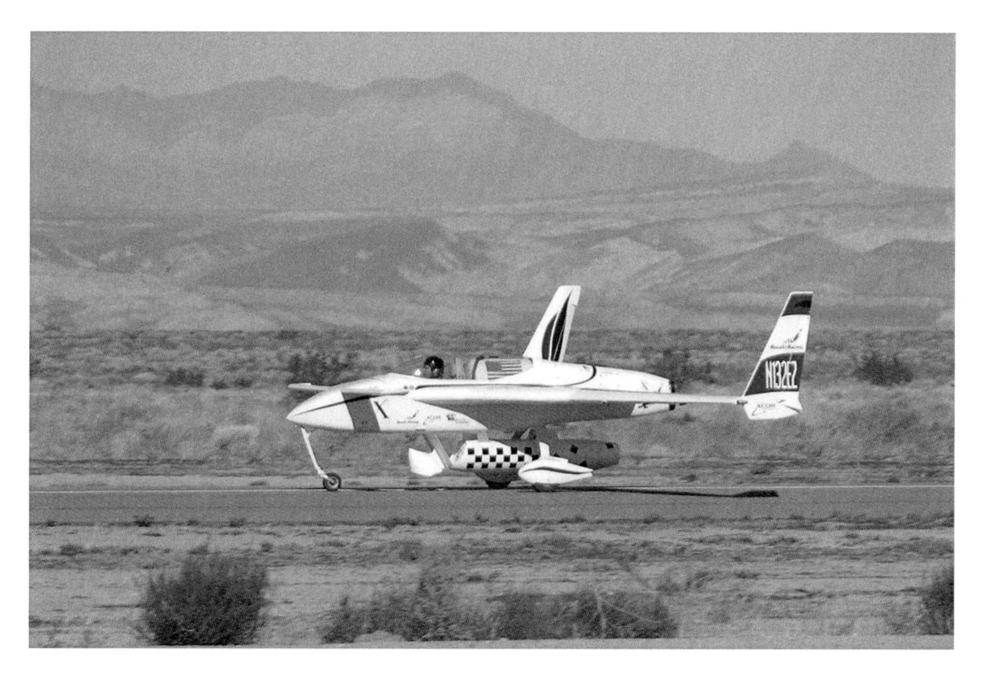

1.2 Rocket racing. Credit: XCOR

[1] Rutan's test flight of the EZ-Rocket marked the first time since the final flight of the X-1 program that a rocket plane had landed after taking off from the ground under its own power.

EZ-Rocket was intended as a test platform for the engine. In its first test in October 2001, Rutan piloted the EZ-Rocket to 2,000 meters in 96 seconds at a speed of 160 knots before shutting down the engines. That 5-minute-20-second flight marked XCOR's first baby step towards suborbital flight.

With the first test flight out of the way, XCOR's focus was on establishing routine and reliable operations first and performance second. Such a philosophy made sense given that most other commercial space programs had failed by prioritizing performance to the detriment of reliability. XCOR was determined not to make that mistake, which is probably one the reasons why the company is still going strong in 2016. Another cornerstone of XCOR's way of doing the business of flying rockets was to use non-toxic fuels, which happen to be a lot cheaper than regular fuels: the EZ-Rocket burned liquid oxygen and isopropyl alcohol, the latter being an item you can find in your medicine cabinet.

While talk at the time was developing the EZ-Rocket in a series of steps that culminated with the building of a supersonic suborbital spaceplane, the design of that vehicle was still undecided, although XCOR thought it might look something like a cross between the Concorde and a Mig-15. The cost? In 2001, XCOR reckoned they could build the vehicle for around US$12 million and that it could be ready in about three years. It was a big dream for such a small company. In 2001, the company comprised just 10 employees, most of whom had worked for the defunct Rotary Rocket company. At XCOR, there was none of the hierarchy and division of duties that are the norm at aerospace behemoths such as Boeing or Lockheed: the shop-floor culture at XCOR was one where each employee was equally responsible for the success or failure of their product. For the most part, it was success and, in 2001, that success translated into a US$300,000 contract from the National Reconnaissance Office to develop rocket motors for use on satellites. Set against the millions of dollars needed to build its suborbital vehicle, US$300,000 wasn't much, but the work was specific to XCOR's goal of refining its development of the rocket motors that would be key to realizing that next-generation vehicle.

Rocket engine development continued into early 2002 with XCOR demonstrating the ability of its rocket motor to shut down and restart in flight. The eighth flight of the series of test flights, which took place in January, was piloted by test pilot Mike Melvill who, in 2004, would go on to find everlasting fame as the pilot of SpaceShipOne. After climbing to 1.5 meters under rocket power, Melvill shut down one engine and then restarted the engine – a procedure he repeated with the second engine before bringing the EZ-Rocket in for a perfect landing.

> "From XCOR's beginning, we have been guided by the vision of rocket operations being as routine as any other form of transportation. Pulling the EZ-Rocket from its hangar, conducting the flight, and towing back to its hangar only took 1 hour, 15 minutes. Today's smooth ground operation and flawless in-flight rocket engine shutdown and restart show that we're getting there."
>
> *XCOR CEO and President Jeff Greason commenting on the success of the EZ-Rocket's eighth test flight*

As testing continued with the EZ-Rocket, XCOR worked on its business case with the result that in July 2002 it was able to announce it had teamed with Space Adventures (see Sidebar) to sell tickets to space. It was a bold partnership given that the vehicle that would be ferrying passengers to the edge of space had yet to be built. But, if all went to plan, XCOR reckoned their Xerus spaceplane (Figure 1.3), capable of carrying one pilot and one passenger, would be ready in three years.

1.3 The Xerus. Credit: XCOR

Space Adventures

Since its founding in 1998, Space Adventures has been the world's leading private spaceflight company, specializing in brokering deals for wealthy individuals to travel to orbit and spend time on board the International Space Station (ISS). Its first ISS client was Dennis Tito, who spent seven days on orbit in April 2001. Since Tito's mission, the company has arranged flights for several other space explorers, including Mark Shuttleworth, Gregory Olsen, Anousheh Ansari, Charles Simonyi, Richard Garriott, Guy Laliberté, and Sarah Brightman.

The partnership between XCOR and Space Adventures was good news for Lance Bass, the N'Sync band member who had failed to attract enough sponsorship money to fund his US$20 million ISS flight. While Space Adventures had not worked with Bass to arrange his flight, it now seemed that the pop star might have another cheaper opportunity to fly to space. At the time of XCOR and Space Adventures, the concept of suborbital travel for the masses was receiving widespread media attention thanks in part to the X-Prize that was offering US$10 million to the first company that could fly a spacecraft above 100 kilometers of altitude and repeat the feat within two weeks. In fact, demand was so great that Space Adventures had collected more than US$2 million in deposits from more than 100 potential suborbital tourists, even though no vehicle existed that could take them to the edge of space.

> "It's at least a multi-hundred million dollar a year business. I don't think there's any technical issues really."
>
> *Eric Anderson, President of Space Adventures, speaking in 2002*

Comments like Anderson's only helped to spur the enthusiasm of imminent joyrides to space. Talk was of a US$1 billion a year travel market, which would have equated to 10,000 commercial astronauts paying US$100,000 every year. Such bold statements seemed to be a little over-optimistic in light of the tragic event just a few months earlier when *Columbia* disintegrated over Texas with the loss of her crew, but perhaps Jeff Greason summed up the collective optimism best when he was interviewed by the *Washington Post* the following year:

> "There are as many reasons as there are people. Why did people settle the West or leave Europe for America? Some wanted to find something. Some wanted to leave something. Some wanted to build something new. There are legitimate differences about the best way to do it. But all of these are good reasons to go into space."

As the Federal Aviation Administration (FAA) and NASA sifted through the wreckage of *Columbia*, XCOR continued its development and testing of the EZ-Rocket, determined as ever to be part of building the future of space transportation. With its stubby little winglets and a chopped-off tail, the EZ-Rocket brings to mind the image of a scaled-down jet aircraft. While it is without a doubt a versatile high-performance vehicle, it doesn't exactly conjure up visions of space access. In the early phase of testing the EZ-Rocket, there were some who argued that perhaps NASA should be developing rocket planes and not some maverick engineers living in the desert. Perhaps, but NASA's recent history of rocket-plane development had been anything but rosy. Take the X-33 (Figure 1.4), for example.The X-33/VentureStar program might have become part of NASA's fleet had it not been for some decidedly iffy decisions during the development of the technology demonstrator. The X-33 was intended to be a prototype for a reusable launch vehicle (RLV) – the VentureStar. To begin with, the development, testing and construction went well, but when it came to the XRS-2200 Linear Aerospike engines, the program encountered a problem that would ultimately bring it to its knees. Despite protests by those working on the program, management had decided to use a composite tank for carrying liquid hydrogen. Project managers were advised that storing liquid hydrogen in a composite tank was an idea doomed to failure, but fabrication went ahead anyway. As predicted, the tank failed during testing. The same happened with a second tank. Faced with imminent project termination, managers authorized the fabrication of the tank using aluminum. This tank proved much easier to build and it was lighter to boot. For a while the project seemed back on track until former NASA director, Ivan Bekey, provided the final hammer blow by insisting the project had to continue with the composite tanks. Since the use of composites would have required a complete redesign of the vehicle, the program was cancelled and US$1.5 billion was lost.

Given the X-33 debacle, one might have been forgiven for thinking that private investors would be scared off by the prospect of developing a rocket plane. After all, if NASA couldn't pull it off, what hope did a small band of engineers living in the desert have? Could XCOR and its competitors do it? Well, the challenge of reaching suborbital altitude was nowhere near as great as reaching orbit. To reach orbit, you need a vehicle capable of attaining a speed of 27,800 kilometers per hour but, to achieve suborbital altitude, that vehicle only has to reach a speed of around 4,500 kilometers per hour, and that's not much faster than some of today's jet fighters (Figure 1.5).So 4,500 kilometers an hour was

1.4 The X-33. Credit: NASA

1.5 Fighter jet. Credit: USAF

XCOR's goal. And the financial incentive? Well, it wasn't the X-Prize because the Xerus didn't meet the three-passenger requirement, but XCOR figured their spacecraft would pay for itself by providing a test platform for research and also by providing joyrides for wealthy tourists. In XCOR's business plan dating back to 2002, the company reckoned development costs would be around US$10 million and that potential annual revenues would exceed US$20 million. That was a bold prediction to make for just one vehicle. But there is a reason that XCOR survived and that is partly because they pursued their goal in a series of small steps. While many of their competitors (Pioneer Rocketplane, for instance) set their collective sights on the grand vision of orbital access, XCOR plodded steadily along, producing hardware and conducting test demonstrations on a budget. One such demonstration took place at EAA AirVenture Oshkosh in 2002, when the EZ-Rocket wowed crowds by performing a series of steep climbs, tight turns, a wingover maneuver, and two mid-flight restarts. It was the perfect demonstration that XCOR knew exactly how to develop and build a reliable, cost-effective, and safe rocket, which happen to be precisely the qualities needed to make access to space routine. Still, despite the aerobatics, it seemed like a long bridge to cross to realizing the Xerus. But XCOR were upbeat as always. In media interviews, they reminded reporters that their suborbital spaceplane wasn't designed to go to orbit so it didn't need to be as robust as the Shuttle had been. And it wasn't as if this was a paper rocket. XCOR had already conducted a fair amount of design work, the engines were being developed thanks in part to the National Reconnaissance Office contract, and they planned to perform hot-fire testing by 2003. The only aspect lagging was the airframe design.

As XCOR continued the development of their Xerus rocket plane, the concept of entrepreneurial space was gaining traction at the highest levels. In July 2003, the Senate Science, Technology, and Space Subcommittee on Space and Aeronautics held a joint hearing entitled Commercial Human Spaceflight. The purpose of the hearing was to discuss the challenges of investing in commercial space ventures and also to examine the regulatory framework. The hearing was an important one for all those companies engaged in the development of suborbital vehicles because the government's role in developing the industry was to create a stable regulatory environment and that in turn could provide incentives in the new market. The hearing discussed the potential market and the way in which the FAA was planning on regulating space tourism. One of the regulatory barriers mentioned was the business of high-altitude flight tests. This, the hearing noted, was subject to FAA safety regulations and those regulations prohibited any company from flying passengers. In 2015, no company has received a license to fly passengers to suborbital space and, in light of the SpaceShipTwo tragedy in October 2014, that license may be a little longer in being approved. The hearing also listened to the concerns of the commercial space entrepreneurs attending, who voiced their concern that the cost of certification could be prohibitive, especially if aircraft certification was to be implemented. The discussion then turned to the question of indemnification and the fallout if one of these suborbital vehicles were to land in a populated area. This issue was important because, in 2002, there was no indemnification regime in place that covered suborbital vehicles. Ultimately, it was decided that vehicle operators were responsible for acquiring liability insurance for their passengers. For XCOR, and all the other operators, the hearing was a key event because regulatory bureaucracy as it pertained to commercial spaceflight was uncharted territory and the industry needed guidelines to work towards licensing their vehicles. One of the biggest headaches was that licensing question because there were two entities of the FAA competing for jurisdiction. Given the impasse between the two branches of the FAA, commercial operators were understandably concerned about the effect such a deadlock could have on investment and project development. XCOR's Jeff Greason was one those invited to present testimony[2] and this is what he had to say on the matter:

> "It's adding a lot of delay and confusion. It's been a problem. I think this argument is without merit. I think the law is already clear. Congress has said suborbital rockets are launch vehicles. We're asking Congress to clarify this once and for all."

Congressman Dana Rohrabacher, Republican chairman of the House Subcommittee on Space and Aeronautics, agreed with XCOR, saying it was unfortunate that the government had not been able to create a regulatory environment. All in all, it was frustrating for the commercial operators, since they had been working with the government since 1999, and after four years the government had yet to provide a definition of what a suborbital rocket was, never mind dealing with all the other regulatory issues. So the commercial operators felt as if they were in limbo, which is effectively what they were until the debate over jurisdiction was settled.

Despite the political viscosity in Congress, XCOR ploughed ahead with its application for a suborbital launch license to the FAA, which it submitted in October 2003. It was the first time any company had submitted such an application and the submission put the FAA on the spot because now the government only had 180 days to rule. If the FAA denied the application, it would be required to report to Congress and provide reasons why the application was denied. The office to which XCOR's application was submitted was the office of the associate administrator for commercial space transportation (AST), which had been transferred from the Department of Transportation to the FAA in 1995. Responsible for licensing commercial rocket launches, it was the AST that now had to consider a variety of factors before deciding whether to award XCOR its launch license. For example, they had to ensure that flying the Xerus didn't pose a threat to public safety and they had to review the environmental impact statement prepared by East Kern Airport District (EKAD), which was the airport from which XCOR proposed to operate.

So, as 2004 rolled around, the FAA was busy drafting a new licensing system that XCOR hoped would clear the way for suborbital passenger flights some time in 2007. For the House Science Committee and the FAA, which were drafting the legislation, it was a difficult task because there were no reference points for training and safety. After all, even the most optimistic commercial space backers accepted that there would be launch failures. The difficult question that members of Congress had to mull over was what constituted acceptable risk? When overseeing the safety of commercial passenger aircraft, the FAA's task is fairly straightforward because there are plenty of rules and regulations governing the certification of these vehicles. A commercial airliner (Figure 1.6) flies at around 800 kilometers per hour at an altitude of about 12,000 meters, whereas a suborbital vehicle will be climbing at a speed in excess of 6,000 kilometers per hour to an altitude of more

[2] Jeff Greason's testimony is included in Appendix I.

1.6 United Airlines Boeing 767-300ER in the Rising Blue livery used from 2004 until the merger with Continental. Credit: Luis Argerich

than 100,000 meters. How was the FAA going to govern that and at what cost to safety? Jeff Greason had an answer and that was to let the passengers decide what was safe, but that opinion wasn't shared by the FAA.In February 2004, Patricia Grace Smith, the FAA's associate AST, told reporters that they were very close to granting a license, although she added that the FAA did not have jurisdiction over establishing safety standards for passengers. But the FAA was able to grant experimental permits and the crew and passengers could sign waivers, relieving the government and the operator of liability in case the flight went pear-shaped. It was also announced that the National Transportation Safety Board (NTSB) would be responsible for investigating spaceship accidents (Figure 1.7).In March 2004, the US Government took a big step towards clearing the way for commercial suborbital flights with the amendment to the Commercial Space Launch Act.[3] For the first time, a suborbital vehicle was defined, and the act stipulated that operators would be required to buy insurance before flying. Among those happy with the amendment was House Science Committee chairman Sherwood Boehlert, who had this to say about the legislation:

"This is about a lot more than 'joy rides' in space, although there's nothing wrong with such an enterprise. This is about the future of the US aerospace industry. As in

[3] The CSLA is a federal law enacted on 30 October 1984. Its purpose was to facilitate commercial spaceflight activities. It is also referred to as the Expendable Launch Vehicle Commercialization Act.

NATIONAL TRANSPORTATION SAFETY BOARD
Public Meeting of July 28, 2015
(Information subject to editing)

In-Flight Breakup During Test Flight
Scaled Composites SpaceShipTwo, N339SS
Near Koehn Dry Lake, California
October 31, 2014

This is a synopsis from the NTSB's report and does not include the Board's rationale for the conclusions, probable cause, and safety recommendations. NTSB staff is currently making final revisions to the report from which the attached conclusions and safety recommendations have been extracted. The final report and pertinent safety recommendation letters will be distributed to recommendation recipients as soon as possible. The attached information is subject to further review and editing.

Executive Summary

On October 31, 2014, at 1007:32 Pacific daylight time, the SpaceShipTwo (SS2) reusable suborbital rocket, N339SS, operated by Scaled Composites LLC (Scaled), broke up into multiple pieces during a rocket-powered test flight and impacted terrain over a 5-mile area near Koehn Dry Lake, California. The pilot received serious injuries, and the copilot received fatal injuries. SS2 was destroyed, and no one on the ground was injured as a result of the falling debris. SS2 had been released from its launch vehicle, WhiteKnightTwo (WK2), N348MS, about 13 seconds before the structural breakup. Scaled was operating SS2 under an experimental permit issued by the Federal Aviation Administration's (FAA) Office of Commercial Space Transportation (AST) according to the provisions of 14 *Code of Federal Regulations* (CFR) Part 437.

1.7 National Transportation Safety Board (NTSB) Executive Summary of SpaceShipTwo accident. Credit: NTSB

most areas of American enterprise, the greatest innovations in aerospace are most likely to come from small entrepreneurs. This is true whether we're talking about launching humans or cargo. And the goal of this bill is to promote robust experimentation, to make sure that entrepreneurs and inventors have the incentives and the capabilities they need to pursue their ideas."

A month later, another piece of history was made when the FAA handed XCOR only the second launch license issued for a manned suborbital vehicle (the first had been awarded to Scaled Composites three weeks earlier). The document was presented to Jeff Greason at the Space Access Conference by George Nield, the FAA's deputy associate AST. Greason, who had only been informed of the presentation the night before, told reporters he was relieved to receive the license, which covered up to 35 flights of XCOR's Sphinx rocket plane.

More media attention was trained on the commercial spaceflight industry six months later when SpaceShipOne (Figure 1.8) took to the skies for the second time in two weeks,

1.8 SpaceShipOne. Credit: D. Ramey Logan

winning the US$10 million X-Prize. The following day, almost every newspaper on the planet had a picture of the achievement that provided a big shot in the arm for the nascent industry.The amendment to the Commercial Space Launch Amendments (CSLA) was eventually stamped legislatively at the end of 2004 following several months of negotiations about the precise role of the FAA in regulating suborbital spaceflight safety. Ultimately, it was decided the FAA would only start regulating the industry if one of the vehicles crashed leading to death and/or serious injury. It was an important victory not only because it empowered the commercial spaceflight industry, but because failure to have enacted the amendment could have set back development of commercial space vehicles such as the Sphinx. In February 2005 (Appendix II), exactly two months after the CSLA amendment was approved, industry leaders met to discuss establishing a Voluntary Personal Spaceflight Industry Consensus Standards Organization with the purpose of developing industry standards to ensure the safety of passengers. Basically, the intent was to go beyond the letter of the law with particular focus on vehicle safety, passenger training (Figure 1.9), and medical certification. The meeting was an acknowledgment of sorts that, while the industry had won the battle for regulation, it still had a long way to go in terms of ensuring crew and passenger safety and, if these issues weren't beefed up, there was a chance the government could step in and do what governments do best: over-regulate. The industry was also mindful that the FAA regulation was limited to 2012, at which point the AST (the very obscure acronym for the FAA's office of commercial spaceflight) could either extend the regulatory period or step in and issue stricter regulations. As part of its job in defining crew and passenger safety, the AST had already begun drafting guidelines on topics such as flight crew operations and medical screening. To its credit, the AST

1.9 The centrifuge: equipment used to train suborbital astronauts. Credit: ESA

had avoided over-using the word "must" in favor of the word "should" in an effort to avoid restricting the industry. As an aside, one interesting guideline pertaining to security was the requirement that passengers should not be allowed to carry weapons on the flight!Perhaps the most detailed of the guidelines drafted by the AST was the medical memorandum, which specified the G-limit that passengers should be subjected to, cabin pressure, and the sorts of medical disorders that passengers should be checked for. Both the medical guidelines and those governing crew operations struck a balance between protecting the passengers and promoting the industry, although there was some contentious debate among those within the FAA about whether the regulations were strict enough. For example, James Oberstar (ranking Democrat on the House Transportation Committee) argued that stricter regulations were required, saying the current regime favored a tombstone mentality. FAA Administrator Marion Blakey disagreed, countering that the administration didn't yet know enough to implement a tougher regime.

> "Do you really think it's a good idea to wait until there's a crash, a fatality, to issue such regulations? Experimentation with human lives, we don't allow that in the laboratories of the Food and Drug Administration or the National Cancer Institute, why should we allow it with space travel?"
>
> *Rep James Oberstar (D-MN), AST conference, Washington, DC, 10–11 February 2005*

At the core of Oberstar's concerns was the question of whether the industry could self-regulate. The argument that the industry can handle the safety of passengers and crew is that it simply can't afford not to because it simply isn't in the interests of any of the commercial spaceflight companies to destroy their vehicles or harm their passengers. One such event could kill the industry. Portentously, Will Whitehorn of Virgin Galactic offered the following sound bite:

"We take safety extremely seriously at Virgin Group, and we wouldn't be entering this industry unless we had a safety culture to bring to it."

All in all, while there were many differences of opinion about whether the industry was over- or under-regulated, the fact that the House Transportation Committee was discussing space transportation was an event in itself, although some were keen to point out that suborbital spaceflight shouldn't be categorized as a mode of transportation, but more of a thrill ride.

With the historic flight of SpaceShipOne and the CSLA amendment, the six-month period between October 2004 and February 2005 had been one of the most eventful in the short history of commercial spaceflight. The outcome of these events also helped the investment case of companies such as XCOR. Following the multi-million-dollar deal with Virgin for a fleet of spacecraft, XCOR had grounds to be optimistic that some of the feel-good factor would rub off on others hoping to invest in the industry – particularly when it came to persuading venture capitalists to take an interest in the nascent industry. And XCOR had grounds to be optimistic because the SpaceShipOne flight had finally demonstrated that it didn't cost hundreds of millions of dollars to build a spaceship. Still, the commercial spaceflight industry was a unique animal because it wasn't the sort of venture that people got into to make a lot of money fast: suborbital spaceflight was definitely in the slow-burner category when it came to return on investment. In 2016 it still is.

Two of XCOR's investors were motivated simply by the dream of affordable spaceflight combined with the frustration that government-funded enterprises seemed to take so damn long. That was the reason that Joe Pistritto invested in XCOR (in June 2000 and November 2000). Pistritto, who made his fortune in the software industry, put money into XCOR after meeting XCOR's founders at a Space Access conference. Another like-minded angel investor is Dr. Lee Valentine, whose motivation for investing in XCOR was to advance his vision of space travel and also make some money. How much money Pistritto and Valentine invested is something of a closely guarded secret because the finances of the suborbital operators – with the exception of Virgin Galactic – have always been rather a hush-hush matter.

As the search for venture capitalist funding continued, the business of rocket racing was being planned as a platform for developing the engines that would one day transport passengers to suborbital altitudes. In October 2005, the Rocket Racing League (RRL) (see Sidebar) was announced. Contestants wouldn't come close to approaching suborbital altitudes, but they would compete on a track in the sky. That "track" would be about three kilometers long and two kilometers wide at an altitude of 1,500 meters. The rocket planes (Figure 1.10) – X-Racers – would take off from a runway and follow a course comprising straights, vertical ascents, and banks – all at speeds of more than 450 kilometers per hour. Think of it like 3D Formula 1.

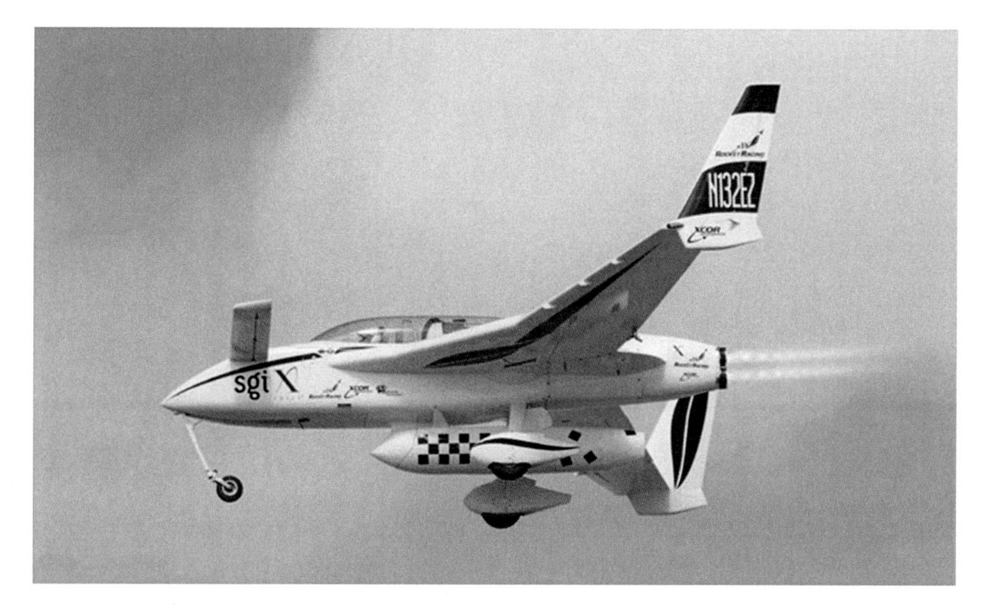

1.10 The EZ-Rocket. Credit: XCOR

Rocket Racing

Aircraft take off in pairs and fly on parallel but separate paths. Every minute or so, another pair of aircraft take off with each pilot following his path in the sky using computer-generated tunnels projected onto the cockpit screen. What the pilot sees is also what the spectators see thanks to giant monitors. Races last for about 90 minutes, which means several pit-stops. Individual flights can be as long as 15 minutes if pilots are skilled at managing their energy. The aircraft are manufactured by Velocity Inc. and weigh about 1,000 kilograms gross.

In 2005, the rules (the league had to be cleared by the FAA) and sponsorship for the league were still being drawn up, but the goal was to hold televised finals in October 2006 and part of the pitch was that this sport could help evolve the development of suborbital spacecraft. And at the heart of the very fast sport was XCOR's liquid-oxygen and kerosene-fueled XR-4K14 rocket engine (Figure 1.11) capable of generating about 1,000 horsepower.

Unfortunately, the RRL, which was supposed to have started in 2006, never got off the ground. No races took place in 2007 either – a year that was remembered for the tragedy at Scaled Composites. The event occurred on 26 July as engineers conducted a flow test of SpaceShipTwo's propulsion system which involved checking how nitrous oxide flowed from a high-pressure tank. It should have been an uneventful test because no engines needed to be fired, but the tank exploded, killing Eric Blackwell, Todd Ivens, and Charles

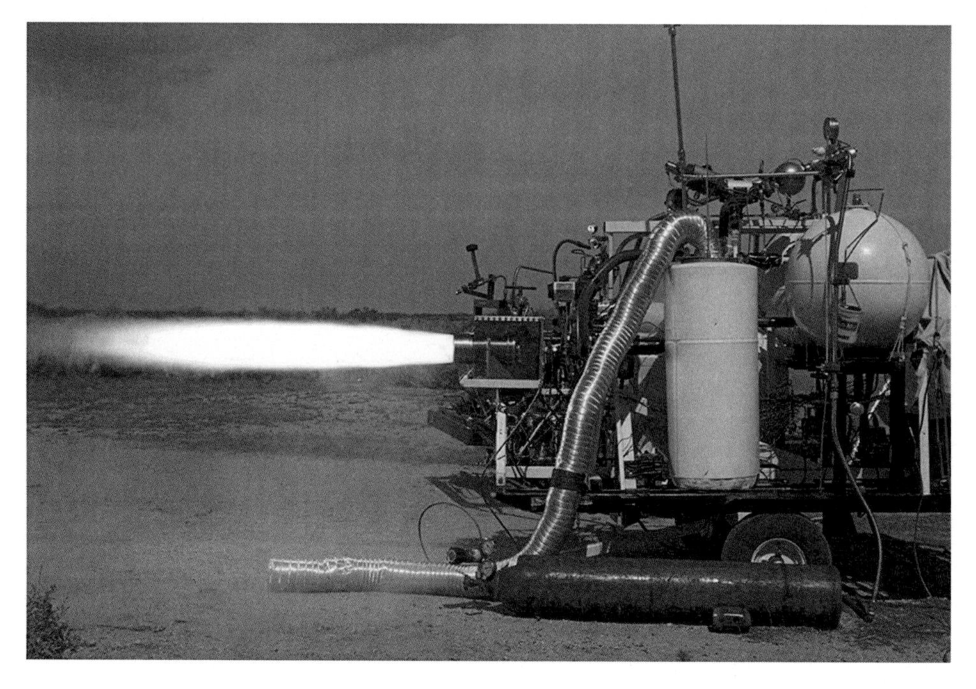

1.11 The XR-4K14 rocket engine being tested. Credit: XCOR

Glen May, and injuring three others. It was a major wake-up call for the industry, although Mojave, a town built on experimental aircraft, took the explosion in its stride:

> "It was inevitable. It was a regrettable thing, but it is a fact of life. There are hazards and risks to building rockets."
>
> *Jeff Greason commenting on the Scaled Composites accident*

In March 2008, XCOR revealed details of the Lynx, the two-seater spacecraft designed to carry one pilot and passenger in the snug confines of a cockpit not much wider than a Cessna 152. Designed to operate several times a day, the vehicle (Figure 1.12) was promoted by the media as being in competition with Virgin Galactic, although this was news to XCOR. Designed to operate like a commercial aircraft, XCOR hoped they would have the Lynx flight-ready by 2010 with revenue flights beginning shortly thereafter. At just 8.5 meters in length, the Lynx is smaller than your average business jet and much smaller than SpaceShipTwo (18.5 meters long), with which the media inevitably made comparisons.The unveiling of the Lynx design came just a couple of months after Virgin Galactic had revealed their design for SpaceShipTwo. Ever since the flight of SpaceShipOne, Virgin Galactic had been the clear leader in the suborbital passenger market but, with the XCOR announcement, the commercial suborbital marketplace had a new vehicle that could potentially compete with SpaceShipTwo. What had happened to the Sphinx and the Xerus? XCOR had continued developing the vehicles but, as they continually adjusted and readjusted the concept and design, the vehicle gradually evolved into a completely different

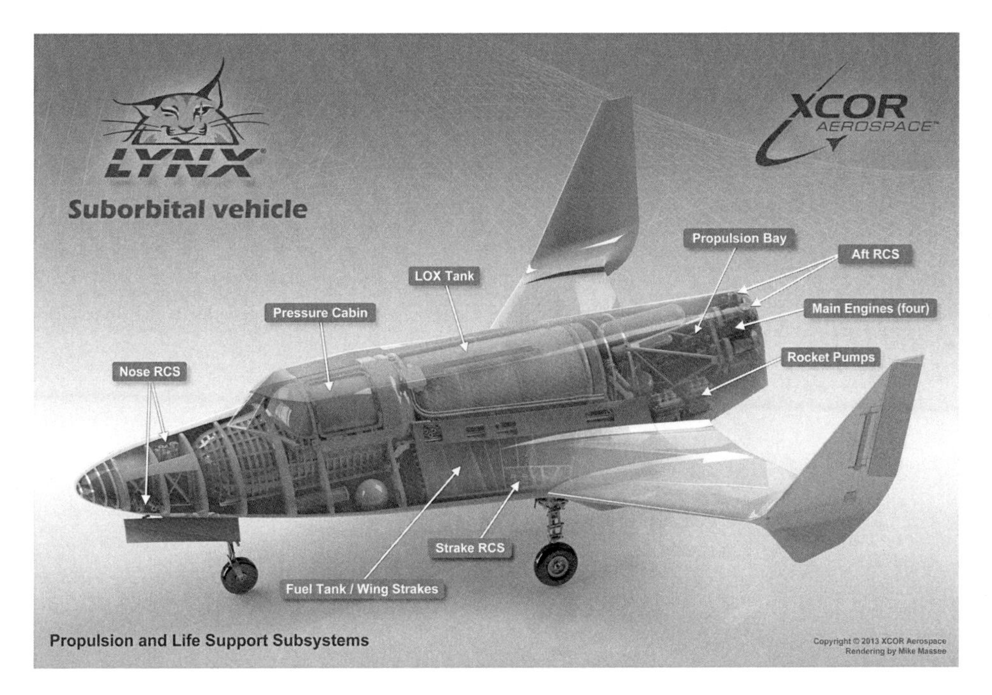

1.12 The Lynx. Credit: XCOR

vehicle and that vehicle was the Lynx. As far as the concept of operations was concerned, the Lynx shared a similar technical approach with the Xerus: a runway take-off, a climb to maximum altitude, and a glide back to the runway. Those with a keen knowledge of space history noted the similarities between the Lynx and the Soviet BOR-4 spaceplane (Figure 1.13), which also happened to be a diminutive vehicle. As of March 2008, XCOR had spent US$7 million developing the Mark I Lynx and they reckoned they would need another US$9 million to complete its development. Once built, the Lynx would be put through a demanding flight-test program by former NASA astronaut Rick Searfoss. Although XCOR didn't put an exact number on the length of the flight testing, Greason said he would be surprised if it took longer than 18 months. Ticket prices were estimated to cost around US$100,000 – a bargain compared to the US$200,000 charged by Virgin Galactic at the time. Having said that, the Mark I was designed to fly to 61,000 meters, which is 39,000 meters short of suborbital altitude. The suborbital-altitude flights would start with the Mark II, which would be developed alongside the flight testing of the Mark I. After all the media attention directed at Virgin Galactic, it was refreshing to see some of that attention diverted, but XCOR wasn't the only company with a suborbital spaceship in development. Rocketplane Global was continuing to develop its XP suborbital vehicle (Figure 1.14), which it hoped to have ready by 2011, and then there was the secret squirrel enterprise known as Blue Origin, which offered very little in the way of details, but was rumored to have a suborbital vehicle in the works.But Rocketplane and Blue Origin were

1.13 BOR-4 spaceplane. Photo by Australian P-3 Naval Reconnaissance Aircraft. Credit: NASA

1.14 Rocketplane Global's XP vehicle. Credit: Space Affairs

slow-burners and the media quickly cooked up a competition between Virgin Galactic and XCOR because it made good press. When asked who was going to be flying passengers first, Greason replied that he was asked that question regularly and he always had the same response, which was that it was nice to be first, but it was better to be right. There wasn't much sensationalism the media could make with quotes like that, but it didn't stop them. The next issue reporters focused on was the difference in ticket prices and they pitched XCOR as the low-fare carrier in a price war that was a product of the media alone and had nothing to do with any statements made by either Virgin Galactic or XCOR. Still, it brought suborbital spaceflight some good press coverage and it may even have helped XCOR gain a customer in the form of the Air Force Research Laboratory, which had decided to use the Lynx to test space hardware in microgravity.

> "Rather than delaying entry into the market until we added all the bells and whistles. The early model is more than sufficient to address a large enough portion of the pent-up demand. Rivalry will drive down prices, and everybody is going to be surprised at how effective real competition will be in benefiting customers."
>
> *Jeff Greason*

At the end of 2008, XCOR had its first travel agent selling tickets on board the Lynx in the form of Arizona-based Rocketship Tours. A deposit of US$20,000 secured passengers a slot in the qualification program and, by the time the venture had been announced, more than 20 would-be passengers had paid for reservations. XCOR's first client was Danish investment banker, Per Wimmer, who, among other lifetime goals, intends to place the Danish flag on the Moon. For those who paid the full ticket price of US$95,000, they could look forward to a five-night stay at the Sanctuary Camelback Mountain Resort and Spa in Arizona, where they would undergo suborbital astronaut training and medical certification. After their stay in the spa, XCORs suborbital wannabes could look forward to another three days in Mojave shortly before their flight for a refresher course. Then, on flight day, after donning a spacesuit and helmet, passengers would climb into the Lynx and rocket up to 61,000 meters. One of those passengers was slated to be Le Roy Gillead who, in February 2009, garnered a free ride as part of XCOR's gambit promoting the space-tourism industry. A World War II pilot and member of the famous Tuskegee Airmen, Gillead was almost 90 when the promotion was announced. It made for great press.

In addition to media coverage about their passengers, XCOR also garnered attention in the military arena. An ambitious plan had been concocted by the US Marines to deliver fast reaction troops anywhere in the world via suborbital vehicles. To that end, Pentagon warfare planners decided to convene a conference and invited representatives from the current crop of commercial spaceflight companies, including XCOR and Virgin Galactic. One of the proposals by XCOR was a plan to provide the Marines with re-entry parachute kits which would allow the super-troopers to bail out of the spacecraft and re-enter the atmosphere before landing under canopy. It was an interesting concept which ultimately went nowhere, which was a pity for XCOR, which needed an infusion of cash. It was late 2009 and the company was searching for a lead investor to put down US$10 million. All the news articles and buzz about suborbital travel and Marines bailing out of spacecraft were great, but money was needed to make it happen. The subject of money was a main topic at the Space Investment Summit held in October 2009 where Greason offered his opinions about the path NASA was steering. Weighing in on the agency's lack of progress, Greason pointed out he was a fan of NASA and that he wouldn't have become a space entrepreneur if he thought the agency was making headway. But pointing to NASA's budget cuts and unaffordable plans such as Constellation, Greason alluded to one of the problems crippling the agency's exploration path: projecting budgets too far into the future without due regard for the economy, policy changes, and political viscosity. He also remarked about the contentious nature of the launch vehicle architecture (at the time, this was the Ares I and V) and noted that to save costs it made sense to fly a smaller launch vehicle more frequently because such a vehicle would have more users and applications. The cogent and economically viable vision of space exploration that Greason described in his speech could have been taken straight out of the XCOR business handbook.

2009 ended on a high note for XCOR with the announcement of its first lease to a South Korean research organization. The US$30 million deal was just another of the many signs of increasing interest in space tourism. Under the agreement with Yecheon Astro Space Center, XCOR would lease and operate one of its Lynx Mark II spacecraft from Korea with an anticipated delivery date of 2012. It sounded like a great deal, but there was the challenge of US export-control laws to deal with, specifically the International Traffic on Arms Regulations (ITAR) which is all-encompassing when it comes to defining dual-use technologies. More good news came in February 2010 when NASA announced it was to invest US$75 million in the Commercial Reusable Suborbital Research (CRuSR) program. The announcement, which was made by NASA Deputy Administrator Lori Garver, at the inaugural Next Generation Suborbital Researchers Conference (NSRC), meant there would be more cash available for flying payloads and that would translate into more funding for companies like XCOR. NASA's plan was to fund each year between 2011 and 2015 to the tune of US$15 million, with the money going to universities and research organizations.

> "I think it's going to shock a lot of people by how transformative it is when access to space becomes like a laboratory instrument, when it becomes something you just go out and do. The immediacy of being able to do science live from space every day of the week is going to be spectacular. Now every researcher takes for granted they'll have one. They don't book a time, they just say 'I need to go do an experiment.'"
>
> *Jeff Greason*

SPACE TRAVEL 2.0

After all the talk of space tourists, the NASA announcement brought the topic of suborbital science and research flights under the spotlight and with good reason because these flights may provide the most significant revenue stream for the industry. The attendees at NSRC certainly thought so. But tourism wasn't out of the headlines for long. Shortly after NSRC, XCOR announced it had signed a US$25 million deal with the government of Curaçao to launch the Lynx as part of the country's plan to establish commercial spaceflight services on the island. Shortly after this announcement, KLM revealed it was to begin offering free suborbital spaceflights for those passengers with high air-mile accounts. The KLM announcement was part of a deal with Space Experience Curaçao, a space travel company founded by Ben Droste and Harry van Hulten, two Dutch pilots turned businessmen. Backed by private equity investors and Curaçao's international airport, the Space Experience Curaçao was promoted as a logical extension of KLM's frequent flier program.

Space Experience Curaçao and XCOR featured in the news again the following year with the announcement that Hensley Meulens, the San Francisco Giants (baseball) hitting coach, was to be launched into space from his native Curaçao sometime in 2014. Keeping with the "celebrity rockets into space" theme, the news that a baseball coach was set to fly into space was followed by the news that Victoria's Secret blond bombshell Doutzen Kroes had also signed up for XCOR's suborbital experience. By mid-2011, XCOR's celebrity passenger list was almost as impressive as Virgin Galactic's, which counted Paris Hilton and Justin Bieber on its manifest. In addition to Meulens and Kroes, those waiting to take the right seat in the Lynx included the world's number-one-ranked DJ Armin van Buuren, founder of Martinair Martin Schröder, and retired swimmer turned Dutch state secretary Erica Terpstra.

While the news of celebrities signing up for suborbital spaceship rides was entertaining press, XCOR still had the business of actually building the Lynx to attend to. By 2011, the company, which had started with just four founders, had grown to number 25, mostly a select group of highly skilled and extraordinarily talented engineers, whose job it was to not only design and construct the spaceship, but also to figure out how it would fly. Up to 2011, there hadn't been much talk about the finer details of how the Lynx would take off, fly to space, and glide back to Earth but, in an interview with the *Daily Kos* in November 2011, Greason revealed some of the specifics of the flight. Passengers could expect to be subjected to loads of up to 4 Gs and the leading edges of the vehicle might reach 260°C during re-entry. During the powered ascent phase, the Lynx would reach a top speed of Mach 2. At an altitude of 42,000 meters and three minutes into the flight, the vehicle's engines would cut off and the Lynx would coast up for another 90 seconds to a maximum altitude of 61,000 meters. The highest G-loads would be experienced during the pull-out phase, which would precede the glide and circle phase prior to landing. Total flight time would be no more than 30 minutes.

THE SUBORBITAL COMMERCIAL SPACE RACE

In 2012, XCOR announced they had completed a key technical milestone by certifying the liquid-oxygen piston pump for the Lynx, thereby clearing the way for integrating the propulsion system into the vehicle's fuselage. The achievement had XCOR suggesting that the first test flights could start at the end of the year, with revenue flights following in 2013. At the time of the milestone announcement, XCOR had grown to 45 employees and the Mark I was taking shape. Thanks to its small size, the vehicle's take-off speed was 190 knots – a speed it could achieve with only 400 meters of runway. Once airborne, the Mark I would pack a powerful punch thanks to its four kerosene and liquid-oxygen engines, each generating 3,000 pounds of thrust. And those liquid-fuel tanks gave the Lynx an advantage over Virgin Galactic's SpaceShipTwo's engine because crews would just need to top off the Lynx's tanks and the vehicle would be ready to go again. SpaceShipTwo's engine would need to be replaced after every flight. Which vehicle would fly first? It was a difficult call to make, although the money was on Virgin Galactic, since SpaceShipTwo's development was more advanced than that of the Lynx. In terms of tickets sold, Virgin had the edge, with more than 500 sold against 225 for the Lynx.

While engineers readied the Lynx, Axe, Unilever's men's grooming brand, decided to promote its latest line of products (Apollo) by announcing it would send 22 of its consumers into space on the Lynx. Contracting with Space Expedition Corporation, Unilever created the Axe Apollo Space Academy (AASA – rhymes with NASA), and revealed that the flights would take off from Curaçao and the winners would be chosen on popular votes (a little different from the selection process that NASA uses to select its astronauts). The company ran a 30-second US$3.8 million advert during the 2013 Super Bowl and then went ahead with a promotional campaign that announced the brand would send 100 finalists to the space camp for a few days of testing and selection. In no time at all, the campaign went viral and suborbital wannabes started marketing themselves on social media (see Chapter 9).

The beginning of 2013 was also notable for the announcement of four experiments that had been sponsored to fly on the Lynx. The sponsoring organization in question was the United States Rocket Academy's Citizens in Space Program, which had cooperated with the Silicon Valley Space Center to acquire a 10-flight contract with XCOR. The 10-flight contract was designed to carry 100 science payloads courtesy of the Cub Carrier (Figure 1.15) and 10 citizen astronauts. It was a bold and unique – as so many ventures in the commercial spaceflight arena are – initiative that not only teamed Citizens in Space with Silicon Valley, but also brought together start-ups such as NanoRack and ArduLab, a microgravity platform. We'll talk some more about Citizens in Space in Chapter 8.Talk of celebrities, Axe-branded astronauts, and rocket academies brought XCOR heaps of media attention, but there was still the business of that spaceship. At the end of 2013, engineers and machinists were still hard at work on test stands working on the 12 engines (the 3N22) due to be integrated into the vehicle. In terms of the commercial suborbital space race cooked up by the media, XCOR was lagging behind because SpaceShipTwo had started its flight-test program in April 2013. SpaceShipTwo had had its own delays, primarily with its propulsion system, which, as in so many aerospace endeavors, was the pacing item that had resulted in predictions of revenue flight being pushed back repeatedly. The sticking point with the Lynx was the carbon-fiber cockpit, but XCOR, with its "we'll fly when we're ready" mantra, didn't seem to perturbed by the successes of SpaceShipTwo.

MIDLAND

As the Lynx continued to take shape (Figure 1.16) in 2014, the FAA announced it had issued a spaceport license to Midland Airport in Texas. This was good news for XCOR because Midland had long been XCOR's relocation destination. The decision to move to West Texas had been taken in 2012 with the plan being to house the corporate headquarters, research and development facilities, and a new hangar at Midland sometime in late 2015, just in time for flight testing the Lynx. By the end of 2014, the summer target date for flight testing seemed achievable. The carry-through spar, which is at the core of the

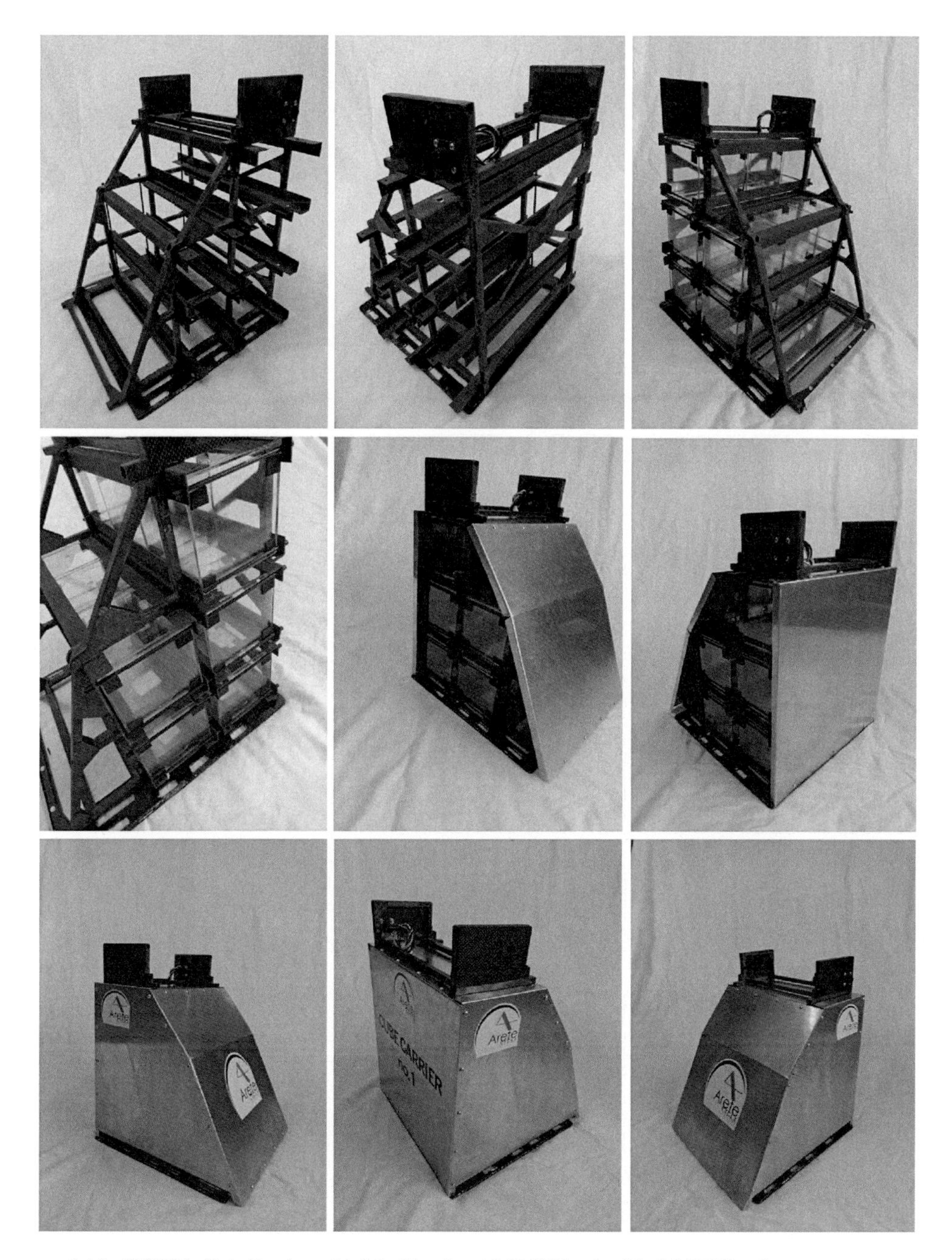

1.15 XCOR's Cub Carrier, which holds nine AMAC Plastics Model 774C polystyrene containers – clear high-density polystyrene boxes that may be modified and joined together to form larger volumes. Credit: XCOR

1.16 View of the Lynx cockpit, fuselage, and strakes. Credit: XCOR Aerospace/Mike Massee

loading structure of any winged vehicle, had been bonded onto the end of the vehicle's fuselage – a step that paved the way for the strakes to be attached. As work proceeded towards that goal in early 2015, XCOR announced the appointment of John Gibson as the company's new CEO and President – a move that allowed Greason to transition to the position of Chief Technology Officer. Shortly after Gibson's appointment, XCOR announced that the strakes had been bonded to the Mark I fuselage – another important milestone that opened the way for integrating subsystems and fitting the landing-gear bays. We'll return to the story of the Lynx's development shortly but, before we do, we need to get the back-story to this remarkable company.

2

Key Players

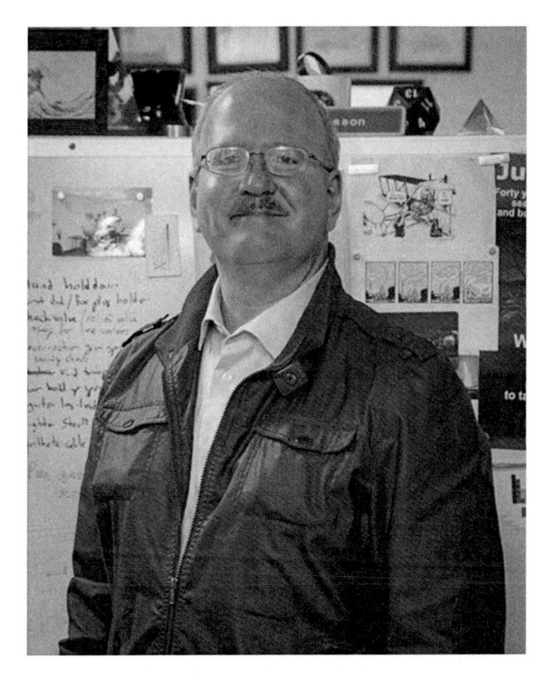

Credit: XCOR

> "This is not optional for me. I believe humanity opening up a frontier in space is important. I think if we want to have a future, we have to do it. As soon as I see other people doing it in a way that I think will work, I can stop. So far, I don't. So I can't."
>
> *Jeff Greason*

E. Seedhouse, *XCOR, Developing the Next Generation Spaceplane*,
Springer Praxis Books, DOI 10.1007/978-3-319-26112-6_2

Rotary Rocket

This was the company that developed the Roton, a reusable single-stage-to-orbit vehicle designed to reduce costs of delivering payloads to orbit by an order of magnitude. Roton's design was as unique and revolutionary as they get. Supported by US$30 million from venture capitalists and angel investors, the Rotary Rocket team, led by Gary Hudson and Bevin McKinney, planned to create a hybrid helicopter-rocket. The idea was that spinning rotor blades powered by jets at the blade tips would lift the spacecraft to an altitude where air density was too thin for helicopter flight. At this altitude, the spacecraft would switch to rocket power. The cone-shaped rocket (Figure 2.1) was designed to bring down the cost of payload to orbit to around US$1,000 per kilogram. While the helicopter-inspired design allowed the Roton to land just about anywhere, the early flight testing wasn't without its problems. To test the hover capabilities, Rotary Rocket built the Atmospheric Test Vehicle (ATV) that flew three test flights: the co-pilot for the tests was Brian Binnie incidentally, who went on to fly with Virgin Galactic (the second X-Prize flight) and then XCOR. The limited visibility in the ATV's cockpit was so restricted that pilots nicknamed it the Batcave. While Rotary Rocket claimed they couldn't continue due to lack of funding, some pointed to unproven technology and a flawed design that led to some unstable landings. Rotary Rocket eventually closed its hangar doors in 2001.

In the revolution that is New Space, there are many key players and personalities. Most people have no doubt heard of Elon Musk and his super-successful SpaceX Falcon rockets and Dragon capsules, just as almost everyone has heard of Virgin Galactic and its flamboyant figurehead, Sir Richard Branson. While these leaders receive the lion's share of media coverage, there are others who pursue the same dream of commercial access to space with a little less fanfare. Take Jeff Greason. This highly experienced engineer may be more reserved than Elon Musk or Richard Branson, but he shares many of the attributes of his more visible counterparts. Like Musk and Branson, Greason has a very clear vision of how to evolve space technology and how to realize a long-term business plan – one that was kick-started way back in 1999 with the founding of XCOR. Before XCOR, Greason worked as an electrical engineer for Intel but, in 1997, he decided to recalibrate his career path by joining Rotary Rocket (see sidebar above) in Mojave. When Rotary Rocket folded two years later, Greason, together with his team of four engineers, decided to strike out and found their own rocket company. Inspired by the X-1 and X-15 programs, they decided XCOR was a catchy name for their rocket company.

Rotary Rocket's demise is just one of many in the short history of commercial spaceflight and it serves as a reminder to those who are still working in the Mojave in the form of a 20-meter-high prototype located in Legacy Park. Greason advises XCOR interns to look but not to touch, reminding them of the local superstition that everyone who has touched the vehicle has lost money. Following Greason (Figure 2.2) were Dan DeLong, Doug Jones, and Aleta Jackson, whom we'll introduce later in this chapter. Without a steady income, the four kept XCOR alive by working on government propulsion contracts.

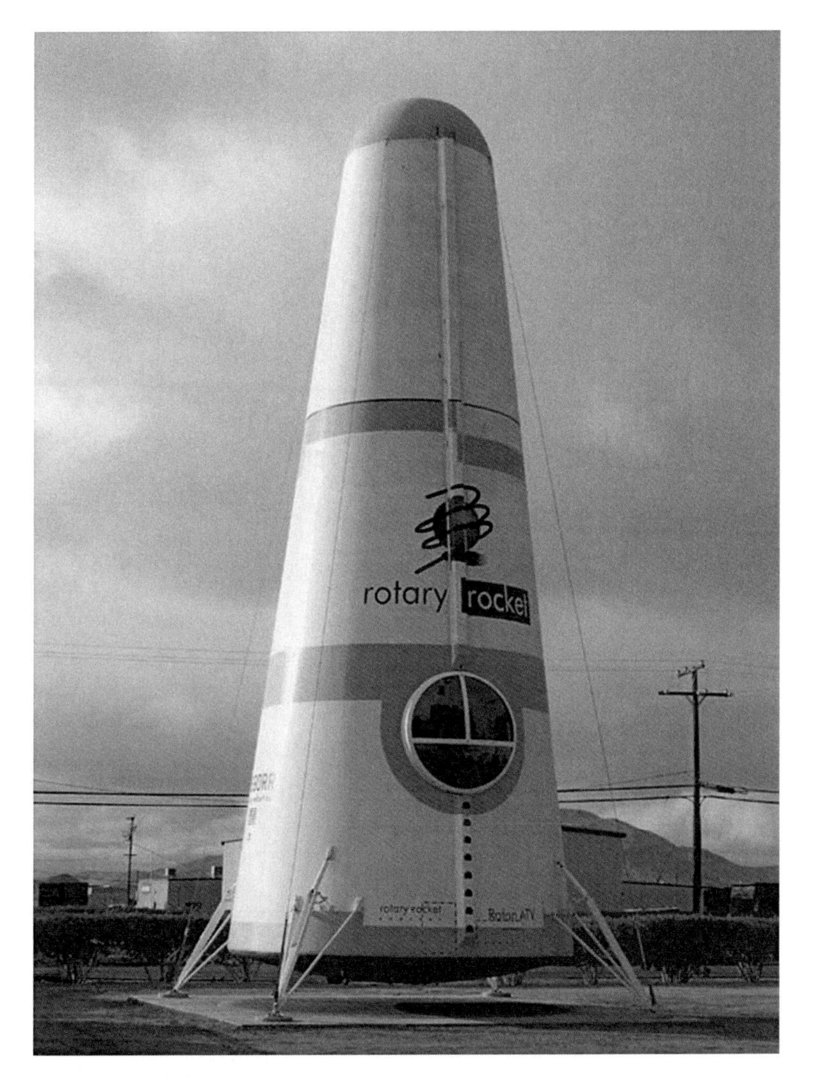

2.1 Rotary Rocket Roton ATV on permanent display at the Mojave Spaceport. Credit: Alan Radecki

JEFF GREASON

I had the very good fortune to meet Jeff at the Next Giant Leap conference in Waikoloa on the Big Island of Hawai'i in November 2014. At the time, my proposal for this book was still taking shape, but I knew it had to feature an interview with the company's Chief Executive Officer, so I asked Jeff if he would mind my asking him a few questions. He graciously gave me 45 minutes of his time and so we sat down in front of the life-sized

2.2 Jeff Greason observing a cold flow test. Credit: XCOR

replica of the Curiosity Rover in the Waikoloa Marriott lobby to discuss XCOR. But, before we get to the interview, here's some background on one of New Space's tech titans.

As a student at the California Institute of Technology, Greason was lucky enough to take a class taught by Richard Feynman. It was Feynman who served on the accident investigation panel following the *Challenger* tragedy, which occurred while Greason was at Caltech. And it was Feynman who garnered most of the media attention with his withering attack on NASA's failure to address critical problems that led to the *Challenger* accident. To Greason, the savaging of NASA by such an esteemed scientist was something of a shock and it got him thinking that perhaps the agency wasn't as untouchable as he had once thought. In fact, Greason reasoned, perhaps government agencies weren't the only ones who could figure out the business of launching rockets into space. It was something he thought about during his career at Intel, where he developed cutting-edge techniques to enable the mass production of new generations of computer chips. His work dramatically reduced the time of development to actually realizing a customer-ready product, which is partly why Intel management awarded him with the prestigious Intel Achievement Award. It was during his time at Intel that Greason attended the Space Access Conference. That was in 1994. After returning from the conference, Greason set about learning everything he could about rockets from assorted rocketry books and engineering journals. Three years later, at the same conference, Gary Hudson, an entrepreneur on the lookout for a technical manager, collared Greason and asked whether he would be interested in building a reusable spacecraft. For Greason, the decision was a slam dunk.

With a promising management career and solid financial security, it must have been tough to walk away into the decidedly hit-or-miss venture that was Rotary Rocket, but Greason had always had his eye on paving the way for civilians to travel in space, so he took the plunge. Fortunately, his wife Carrine (whom he met while at college in Portland) supported him, reasoning it was best to have a happy spouse and viewing the move from

Portland, Oregon, to the Mojave as an adventure (thanks to her freelance job as marketing communications support to high-tech companies, she could work wherever there was an internet connection). It helped that Greason had a supremely talented engineering team working for him at Rotary and the fact he had squirreled away a decent nest egg while working at Intel. After Rotary folded, Greason's income rolled down to almost zero, so he siphoned off funds from his Intel stock while making the commute from Tehachapi to the flight line at Mojave (for the first year, to keep the company going, Greason's engineering team used credit cards to buy parts). With quiet streets, few stores, and the odd signal light, Tehachapi, with a population that hovers around 30,000, is a world away from the big-box towns, as is Waikoloa, which is where the following interview took place.

One of the first questions I had was when test flights would commence.[1] Since the 2004 flight of SpaceShipOne, the commercial spaceflight industry seems to have been in a perpetual holding pattern – one which seemed to have stalled with the SpaceShipTwo tragedy that took place less than two weeks before I interviewed Jeff. The plan was to start the flight-test program in late summer 2015, with the aim to fly as many as 80 test flights of the Mark I. When asked how long he expected this would take, Jeff replied he would be surprised if the test program could be completed in less than six months and doubted it would take longer than 18 months. But, he added, the Lynx would fly when it's ready to fly and XCOR's engineers were working as fast as they could to make that happen. The last thing XCOR wanted to do was to apply extra pressure by announcing a flight date. For those of you who have followed Virgin Galactic, you may remember the seemingly never-ending pronouncements of dates for when revenue flights would start. First it was 2009 and then 2010. Then the absolute latest date for rocketing passengers into suborbital space was 2013. Following the SpaceShipTwo accident, Virgin Galactic has adopted the XCOR mantra. A case in point: when asked by a member of the audience at the 2015 Space Access Society conference in Phoenix when SpaceShipTwo would fly, Will Pomerantz followed the XCOR mantra and replied it would fly when it was ready. And, before flight testing could begin, Jeff reminded me, there were still a few tasks that needed to be completed, the first of which was to piece together all the structural subassemblies (Figure 2.3) and after that they had to begin debugging the propulsion system. And, while he couldn't predict when all these pieces would come together, Jeff was confident that work was progressing as it should.

Next, I asked for Jeff's perspective on the competition in the commercial suborbital and orbital spaceflight arena. Of all the people in the business of New Space, Jeff is recognized as one of the most knowledgeable. He is an expert in the Federal Aviation Administration (FAA)'s office of commercial spaceflight (AST) reusable launch regulations and is a co-founder and vice chairman of the Personal Spaceflight Federation so, when he offers his opinions on the subject of where commercial spaceflight is heading, people sit up and take notice. Jeff is convinced the suborbital and orbital markets will remain very distinct entities for quite some while for the simple reason that there are no overlapping segments. On the subject of the orbital market, he thinks it will be dominated for quite some time by expendable launch vehicles and that it may take some time before reusable vehicles take

---

[1] Rather than laying out the typical question-and-answer format, I have condensed Jeff's replies into the following narrative.

2.3 Reaction Engines' Skylon. Credit: Reaction Engines

center stage. But, when reusable vehicles come on stream and can be flown regularly, they may come to dominate the lower end of the market. On the subject of the suborbital market, he is very confident that XCOR will be in a very strong position to compete on price because the company's capital costs are so low and the design of the Lynx means that the company can fly up to four flights per day: that won't be the case with Virgin Galactic, although SpaceShipTwo can carry up to six times as many passengers than the Lynx.

On the subject of point-to-point transportation (Figure 2.4), Jeff recognizes that there is a lot of interest in this mode of high-speed travel, but he doesn't expect it to be realized before the mid-2020s. One of the reasons for this, apart from the tremendous technological challenges, is the competition with regular commercial air travel. The market for those who need to get somewhere very, very fast while paying stratospheric ticket prices is very small. And, despite all the nice pictures in the glossy magazines, this mode of travel isn't exactly practical because any prospective passenger first has to travel to the remote spaceport to catch a flight which will land at another remote spaceport. At the end of the day, this ultra-fast mode of transportation may not end up being that fast at all. And, as far as using the Lynx for point-to-point travel is concerned, the idea is a non-starter because the vehicle can't fly much more than 320 kilometers downrange – a distance that could be extended by subsonic glide, but not enough to make point-to-point travel financially viable.

Another sensitive issue we discussed was the International Traffic on Arms Regulations, or ITAR. Part of XCOR's business plan is to launch the Lynx from other countries, but to do that they need to transport the vehicle outside of the US. The problem, according to Category XV (Spacecraft and Related Articles) of the US Munitions List (USML), is that

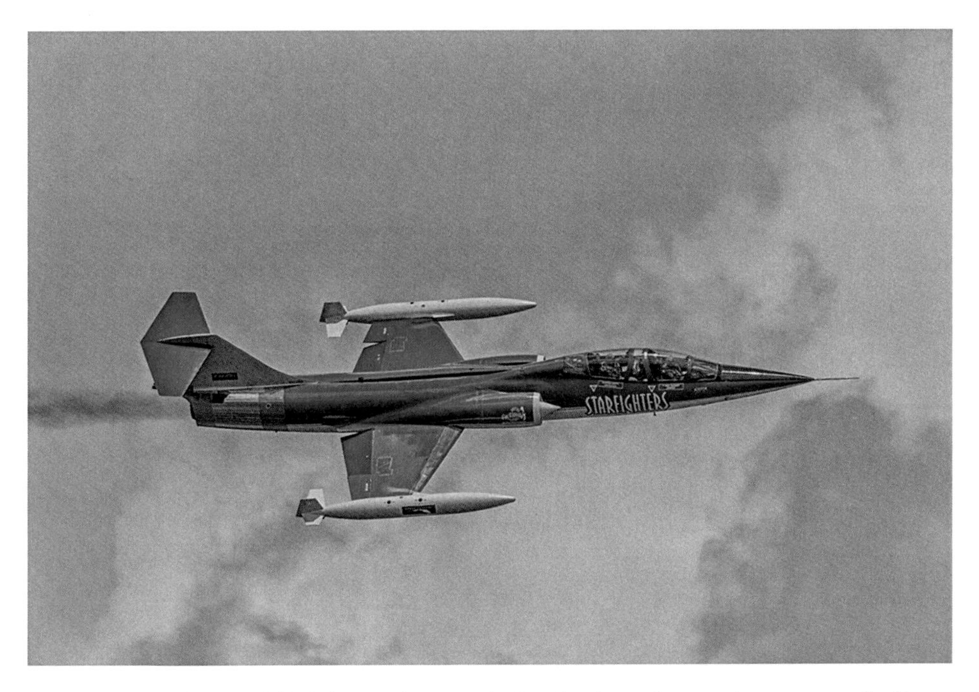

2.4 The Starfighter 104 is being used as a test bed to develop point-to-point transport. Credit: Starfighter

the US Government considers tanks, fighter jets, bombs and … suborbital spacecraft as munitions. That's right: suborbital vehicles are a regulated item, right along with ballistic missiles. In the 2000s, this classification caused more than a little consternation among those in the commercial spaceflight arena and ITAR quickly became a four-letter word. It still is. In May 2014, the State Department revised Category XV and removed some commercial satellites and components used to build those satellites, but the hot-button issue of suborbital vehicles remained. The rationale was cited as follows:

> "For example, launching spacecraft to sub-orbit or orbit requires MTCR Category I items, upon which are placed the greatest restraint with regard to export. Spacecraft specially designed for human space flight that have integrated propulsion present another security concern, for such capabilities may be used for the purposes of weapons targeting from space. So, although these technologies and capabilities are used in commercial endeavors, they continue to merit control on the USML."
>
> The State Department's interim final rule for the revised Category XV of the US Munitions List.

While the satellite manufacturers were reasonably happy with the amendment, the commercial spaceflight industry was less than impressed. There had been some hope that the State Department would have moved suborbital vehicles to the less restrictive Commerce Control List (CCL), but it didn't, which is something that frustrates Jeff because it not only means that XCOR is banned from transporting the Lynx out of the country, but also prevents the company from hiring non-US nationals.

DAN DELONG

Dan DeLong (Figure 2.5) is XCOR's Vice President and Chief Engineer. His career as an engineer got started while working as an underwater equipment designer for Westinghouse Ocean Research and Engineering, where he developed emergency life-support equipment and worked on closed-circuit breathing gear. From Westinghouse, DeLong moved to Perry Oceanographics, where he worked as the company's Staff Materials Engineer between 1978 and 1983, during which time he helped develop the company's remotely piloted vehicle RECON III. After Perry, DeLong spent five years developing air-launched orbital vehicles while employed by Teledyne Brown Engineering, working alongside famed Dr. Ernst Stuhlinger who had been brought to the US as part of Operation Paperclip along with Wernher von Braun. During his time at Boeing, DeLong also served as an analyst for developing the International Space Station (ISS)'s life-support system. For a couple of years after his stint with Boeing, DeLong was president of Eureka design, which built hardware for Kistler Aerospace, but, in 1997, he moved on and co-founded Rotary Rocket. When that adventure ended in 1999, he followed Jeff and co-founded XCOR. He's stayed there until November 2015 before joining Jeff to found Agile Aero.

To say that DeLong is one of the world's leading rocket propulsion innovators is an understatement. In 2002, *Esquire* magazine featured him in one of *America's "Best and Brightest" 43 People Who Will Revolutionize the World.* Which is exactly what DeLong has been doing in the Mojave for the past 15 years. If you're planning on flying on the Lynx, you'll be taking a ride in a vehicle that has been designed by one of the very best engineers in the business. And that's partly because the Lynx isn't the first reusable launch system DeLong has had a hand in designing. While working for Teledyne, he worked on the Spaceplane and the Frequent Flyer, both air launched reusable vehicles. The Spaceplane

2.5 Dan DeLong. Credit: XCOR

was designed to be mounted on a converted 747 Carrier aircraft and launch up to three tonnes of payload to 400-kilometer low Earth orbit (LEO). The idea was that it would be built with off-the-shelf components and be powered by one Space Shuttle Main Engine (SSME) and six RL-10s. After delivering its payload to LEO, the winged single-stage-to-orbit (SSTO – with the assistance of the 747) vehicle would glide back for a runway landing. Although the Spaceplane wasn't a thoroughbred SSTO since it benefitted from the air launch from the 747, the mission design was elegant in the way it solved the challenges of a pure SSTO vehicle. To design a ground-launched vehicle capable of horizontal take-off and landing is still a significant challenge, even in the mid-2010s. That's because a ground-launched vehicle needs to be fitted with landing gear that supports the full weight of the vehicle and wings that must be capable of producing lift at the very low take-off speeds. As if that isn't bad enough, the vehicle must have engines that can operate equally well at sea level as in a vacuum. How do you resolve these problems? DeLong decided to use air launch – a decision that made the whole flight much, much easier. For one thing, there are fewer meteorological uncertainties at higher altitudes, which means that fuel reserves can be reduced. And, since the launch occurs at high altitude, this means that not only are aerodynamic drag losses less, but Max Q is less also, which lowers the structural mass of the vehicle. Also, because the vehicle is no longer required to lift the full weight at low take-off speed, the wing area can be reduced, which further reduces structural mass. Finally, the mission flexibility of an air-launched system is much greater than a thoroughbred SSTO because the carrier aircraft can fly up-range if necessary (to optimize the launch point relative to an orbital destination, for example) and this ability also permits a greater return-to-launch-site abort window. In addition to the Spaceplane, DeLong also worked on the Frequent Flyer, an unmanned vehicle that was also designed to be launched from a 747. The Frequent Flyer's job was to deliver 300–450-kilogram satellites to LEO, although there was an option to carry passengers using a special pod. Unfortunately, neither the Spaceplane nor the Frequent Flyer was built, but that isn't the case with DeLong's current project.

DOUG JONES

As XCOR's Chief Test Engineer, it is Doug's job to deal with the test design and analysis of testing of the Lynx's engine development – a job description that fits his skill-sets like a glove. Born with an unusual ability to pinpoint the most minute of minute propulsion anomalies, his colleagues long since conferred the title of the Rocket Whisperer upon him. Prior to joining XCOR, Doug, like Dan and Jeff, worked for Rotary Rocket, where he was tasked with coordinating the design of the rocket engine and interpreting the reams of test data. As part of his job as flight-test engineer, Doug flew the X-Racer on a number of occasions. Ensconced in the right seat, Doug's job was to keep an eye on the propulsion system as the pilot put the aircraft through its paces. This job was achieved thanks to the strategic positioning of myriad sensing devices in the engine and a mini camera (one of four) attached to the vertical stabilizer that was focused on the engine exhaust. During the 37-flight X-Racer program, Doug worked with Primary Flight Test Engineers Mark Street and Douglas Jones, to troubleshoot the detail and system design choices affecting the performance of the aircraft. This process was a relatively quick one thanks to the rapid turnaround

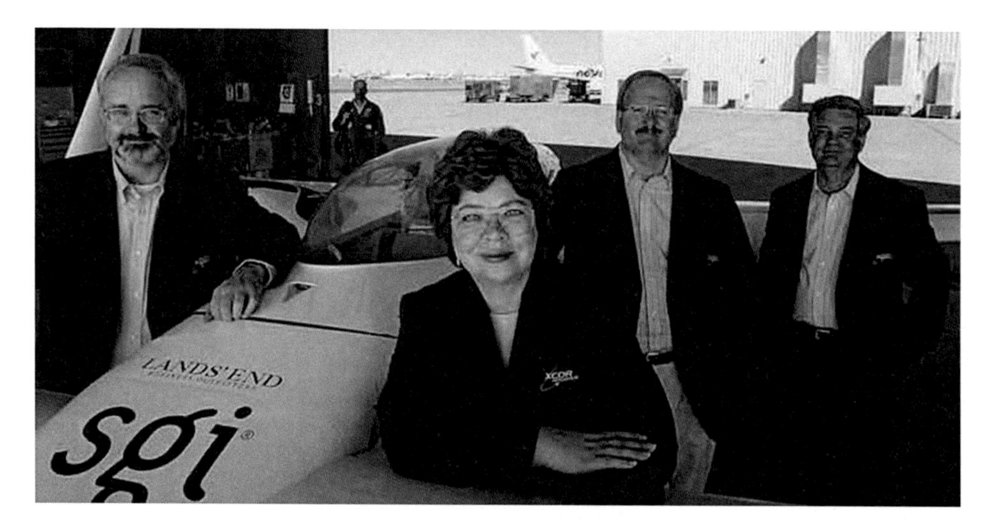

2.6 Aleta Jackson. Credit: XCOR

between flights. During the flight, the engineer would identify the problems, the issues would be fixed on the ground, and in short order the aircraft would be flown again.

ALETA JACKSON

A technician, editor, and pilot, Aleta (Figure 2.6) is XCOR's Chief Technician and Office Manager. Like Jeff, Dan, and Doug, her previous employer was Rotary Rocket where she managed technical documentation. Prior to her stint at Rotary Rocket, she worked for McDonnell Douglas and Electron Emissions Systems, although she is perhaps best known as the first woman to fly the X-Racer.

THE TEST PILOTS (FIGURE 2.7)

NASA astronaut Rick Searfoss

As momentum has continued to build in the commercial spaceflight industry, it has opened up a second career for some spacefarers leaving the astronaut ranks. For example, in July 2014, Bigelow Aerospace hired former NASA astronauts Ken Ham and George Zamka. Zamka, who had left NASA in 2013 to work at the FAA's Office of Commercial Space Transportation, was an ex-military pilot who flew on Shuttle missions STS-120 and STS-130. Ham, who joined Bigelow from his job as chairman of the US Naval Academy's Aerospace Engineering Department, was a US Navy Captain who flew on STS-124 and STS-132. One of his jobs at Bigelow Aerospace will be to develop astronaut training programs for Bigelow's sovereign customers who will be spending time on the company's orbiting habitats. SpaceX is another New Space company who had hired NASA retirees, counting Garrett Reisman among their employees. Reisman, who is the project lead for the

2.7 Rick Searfoss. Credit: NASA

manned Dragon variant, the V2, flew with Ham on Shuttle flights STS-124 and STS-132 in addition to flying on the *Endeavor* during STS-123, which was the flight that delivered him to the ISS as a member of Expedition 16. While SpaceX and Bigelow Aerospace attract more media attention than XCOR, the trend of employing NASA's retired astronauts was one that was started by the Mojave-based company when it hired Searfoss to fly the X-Racer in 2008.

In Searfoss, XCOR was one of the most accomplished pilots in the US Air Force (USAF) and an astronaut to boot. A Distinguished Graduate of the USAF's Top Gun School and the Naval Test Pilot School, Searfoss has accumulated more than 6,000 hours of flight time in more than 70 types of aircraft, including the X-Racer. It was while he was working as an instructor at the test pilot school that Searfoss was selected by NASA for its astronaut program. After graduating as an astronaut in 1991, Searfoss didn't have to wait long before being assigned to and flying a mission. His first was STS-58 (Figure 2.8), which launched on 18 October 1993 – a flight on which he piloted *Columbia*.

STS-58 was followed in short succession by STS-76 (22–31 March 1996) and STS-90 (17 April 17 to 3 May 1998) – a flight for which Searfoss served as Commander. All told, Searfoss had logged 39 days in space by the time he retired from NASA in 2003. And, for a deposit of just US$20,000, you can book a seat next to Searfoss on an upcoming Lynx flight.

Commercial astronaut Brian Binnie

Like Searfoss, Binnie is a supremely experienced pilot with a stellar resume. A graduate of the US Navy's Test Pilot School, Binnie has accumulated more than 5,000 hours of flight time on more than 60 aircraft types and has an Airline Transport Pilot's license to

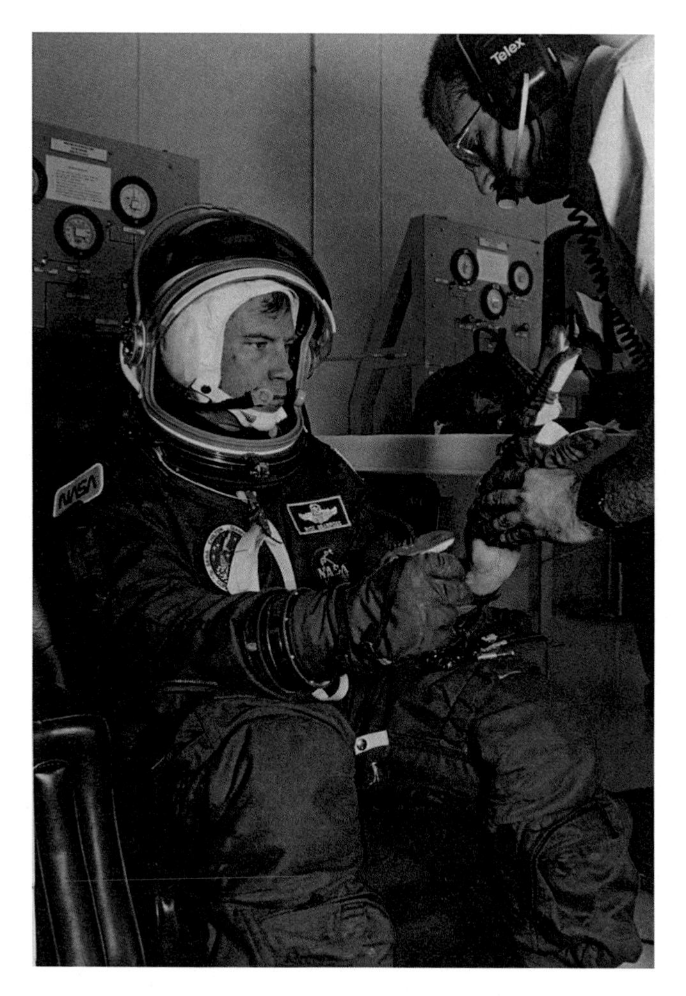

2.8 Rick Searfoss. Credit: Rick Searfoss

boot. Much of his test-flight experience was logged flying the F/A-18 while expanding the launch envelopes of various weapons systems and testing the transonic performance of the aircraft. In addition to his military experience, Binnie also gained skills in the commercial space sector as a test pilot for the Roton venture (where he worked alongside Greason and DeLong) – a program for which he developed the aircrew checklists and emergency procedures. After Roton folded, Binnie headed for Virgin Galactic, where he found everlasting fame when he piloted SpaceShipOne on the second X-Prize flight (Figure 2.9) – a flight that earned him his commercial astronaut wings.

> "I wake up every morning and thank God I live in a country where all of this is possible. Where you have the Yankee ingenuity to roll up your sleeves, get a band of people who believe in something and go for it and make it happen. It doesn't happen anywhere else."
>
> *Brian Binnie, after piloting SpaceShipOne on 4 October 2004*

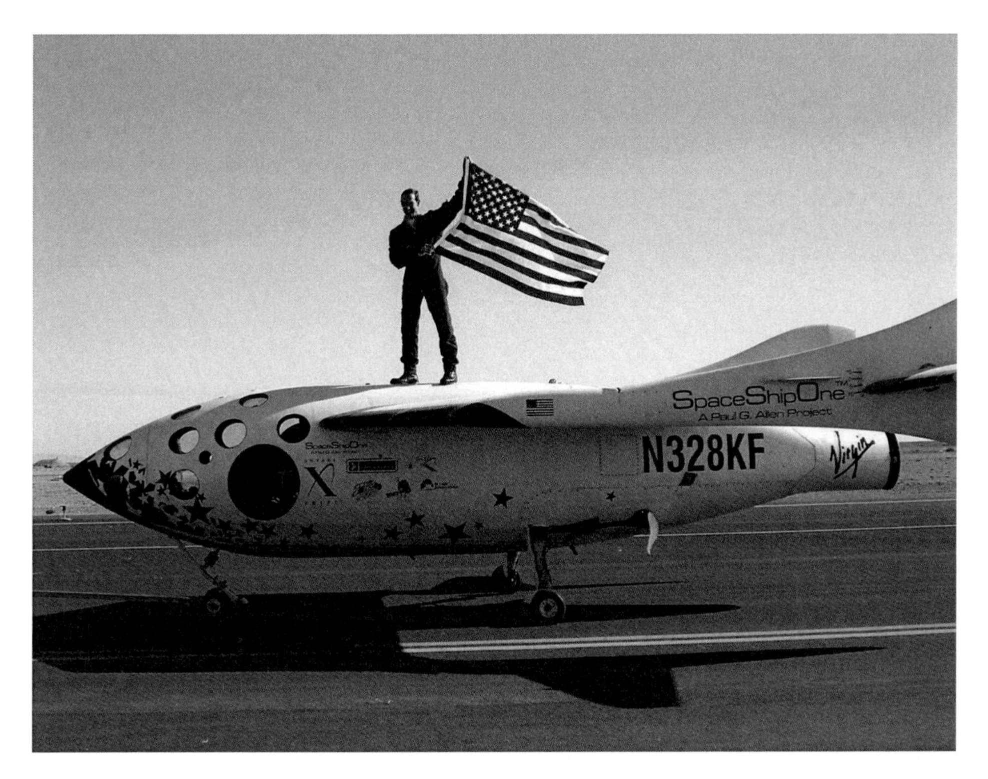

2.9 Brian Binnie. Credit: D. Ramey Logan

On 9 April 2014, almost 10 years after that historic flight, Binnie was in the headlines again when XCOR announced it had hired the distinguished pilot as the company's senior test pilot.

> "The combination of Rick Searfoss and Brian Binnie at XCOR is a powerful statement from the professional flight test community about XCOR and the Lynx. The pairing of two people who are decorated military test pilots, rocket-powered aircraft pilots and astronauts is a powerful team that defines XCOR as a leader in the industry."
>
> *Andrew Nelson, commenting on XCOR's hiring Brian Binnie*

3

Industry

Credit: XCOR

Before we look at the design and development of the Lynx, it's helpful to understand the industry in which the vehicle will operate. As it stands in 2016, there is no suborbital passenger industry because the very few operators out there building suborbital spacecraft still have a way to go before they can start revenue flights to suborbital altitudes. When those revenue flights may start is hard to say because operators are loathe to commit to a date. Before the SpaceShipTwo accident of October 2014, Virgin Galactic made pronouncements every few months about when passengers would be flying into space. All those predictions came to nothing and, after the SpaceShipTwo accident, Virgin's mantra has been that they will fly when they're good and ready. But, when the day finally rolls around on which the Lynx starts flying paying passengers, whether they be thrill-seeking tourists or serious scientists, how robust will the industry be and what bureaucratic machinations will affect that industry? We don't know the answer to the first question because the industry has yet to take off, but an organization by the name of the Tauri Group has crunched some numbers, done some crystal-balling, and come up with what it thinks is a

E. Seedhouse, *XCOR, Developing the Next Generation Spaceplane*,
Springer Praxis Books, DOI 10.1007/978-3-319-26112-6_3

reasonable 10-year over-the-horizon forecast. We'll take a look at the Tauri Group's forecast first. In terms of administrative maneuverings that will certainly affect the industry, there exist the International Traffic on Arms Regulations (ITAR), the Federal Aviation Administration (FAA), and the United States Munitions List (USML) (yes, that suborbital spacecraft you may be planning to take a trip on is classified as a weapon), which will also be discussed.

THE TAURI GROUP

The Tauri Group promotes itself as an innovator in analytical consulting – a job that it does by applying imaginative and creative data-driven analysis to a particular subject, whether that be an investment strategy for the satellite industry, a strategic analysis for some aspect of national security, or, in this case, the imminent suborbital spaceflight market. The group has no vested interests and has a long history of providing governments and companies with sound and accurate industry analysis and market assessments. In 2012, the Tauri Group's Space and Technology team released their assessment of the future of suborbital reusable vehicles in a report entitled *Suborbital Reusable Vehicles: A Ten-Year Forecast of Market Demand.*[1] The US$277,000 study was paid for by the FAA and Space Florida, which is the state's spaceport authority and space development organization. The FAA is the government entity that regulates and licenses the US commercial launch industry, which means it is also responsible not only for ensuring public health and safety, but also for protecting national security and foreign policy interests during launches such as a suborbital passenger launch. Space Florida, meanwhile, is the state's spaceport authority and space development organization, which means it is committed to attracting the next generation of space business. So both the FAA and Space Florida were keen to find out on what kind of trajectory this suborbital space business was headed – hence the Tauri Group's report.

The full report is 102 pages long and I've provided a link for those who are interested in reading these documents cover to cover. For those who prefer the nuts and bolts, what follows is a synopsis. To begin with, the Tauri Group divided the demand for suborbital reusable vehicles (SRVs) into eight markets (Figure 3.1), it then presented three growth scenarios (a *baseline*, *growth*, and *constrained* scenario), and finally it differentiated between individual and enterprise user communities.

> "Our study concluded that demand for SRV flights at current prices is genuine, sustained, and appears sufficient to support multiple providers."
>
> *Ms. Carissa Christensen, managing partner, Tauri Group*

Given that the subject of the study has yet to find its wings, it's not surprising that the report includes a number of caveats concerning the assumptions the Tauri Group used when conducting their research. For example, the study includes a section with the title "Major Uncertainties," which describes the unpredictable nature of making some of the

---

[1] You can read the full report at *www.nss.org/transportation/Suborbital_Reusable_Vehicles_A_10_Year_Forecast_of_Market_Demand.pdf.* For more information, you can visit their website: *www.taurigroup.com.*

COMMERCIAL HUMAN SPACEFLIGHT
Human spaceflight experiences for tourism or training
Individuals
Corporate
Contests and promotions
In-space personnel training

BASIC AND APPLIED RESEARCH
Basic and applied research in a number of disciplines, leveraging the unique properties of and access to the space environment and microgravity
Biological and physical research
Earth science
space science
Human research

AEROSPACE TECHNOLOGY TEST AND DEMONSTRATION
Aerospace engineering to advance technology maturity or achieve space demonstration, qualification, or certification
Demonstrations requiring space/launch environment
Hardware qualification and test

MEDIA AND PUBLIC RELATIONS
Using space to promote products, increase brand awareness, or film space-related content
Film and television
Media, advertising, and sponsorship
Public relations and outreach
Space novelties and memorabilia

EDUCATION
providing opportunities to K-12 schools, colleges, and universities to increase access to and awareness of space
K-12 education
University educational missions

SATELLITE DEPLOYMENT
The use of SRVs to launch small payloads into orbit
Very small satellite launch

REMOTE SENSING
Acquisition of imagery of the Earth and Earth systems for commercial, civil government, or military applications
Commercial Earth imagery
Civil Earth imagery
Military surveillance

POINT-TO-POINT TRANSPORTATION
Future transportation of cargoor humans between different locations
Fast package delivery
High-speed passenger transportation (civil)
High-speed troop transportation (military)

3.1 The Tauri Group. Credit: Tauri Group

predictions laid out in the report. In terms of actual forecasting, the study sticks to hard numbers with regard to actual seats or cargo equivalents, and also an indication of the revenue that suborbital operators might reasonably expect for each growth scenario. For instance, for its baseline scenario, the report predicts 370 seats in the first 12 months of operations – a figure that grows to 500 seats per year in the tenth year, and a forecast that results in 4,518 seats (or seat equivalents) in 10 years (Table 3.1). That's across the whole

Table 3.1 Tauri Group's forecasted demand for SRVs by seat/cargo equivalents[1]

	Year										
Scenario	*1*	*2*	*3*	*4*	*5*	*6*	*7*	*8*	*9*	*10*	*Total*
Baseline	373	390	404	421	438	451	489	501	517	533	4,518
Growth	1,096	1,127	1,169	1,223	1,260	1,299	1,394	1,445	1,529	1,592	13,134
Constrained	213	226	232	229	239	243	241	247	252	255	2,378

[1]Full report available online at *www.nss.org/transportation/Suborbital_Reusable_Vehicles_A_10_Year_Forecast_of_Market_Demand.pdf*. For more information, see the group's website at *www.taurigroup.com*.

industry, not just one operator, incidentally. For that same scenario, the report calculates total revenue of US$600 million, which compares to a figure of US$1.6 billion in the growth scenario and just US$300 million in the constrained scenario (such a scenario might occur as a result of a financial crisis or a fatal accident). If the industry goes swimmingly, then the growth scenario predicts a whopping 1,592 seats by the tenth year. That should keep XCOR and Virgin Galactic happy.

In addition to forecasting seats and revenue, the report also takes a look at the companies that are in the process of developing SRVs, including Armadillo Aerospace, Masten Space Systems, Blue Origin, Virgin Galactic, and of course XCOR. Some of these companies, such as Masten, aren't planning on launching passengers anytime soon, whereas others such as Blue Origin are planning manned operations, but have a way to go in the development of their vehicle. And, based on the Tauri Report, those companies probably need to fly passengers sooner rather than later because the dominant SRV market is the human one by some margin: Tauri reckon manned suborbital flights will represent 80% of total demand. To arrive at that conclusion, Tauri identified 8,000 high-net worth individuals (those whose worth was US$5 million or more) from around the world who were interested in buying a suborbital ticket at the prices advertised in 2012. About one-third of these wealthy individuals were from the US. Tauri predicted that 3,600 of the individuals, or 40% of the total, would fly a suborbital flight within 10 years. At the time of the report, 925 people had actually bought tickets. As for the remaining 20%, Tauri predicted Basic and Applied Research would account for 10% of demand and the final 10% slice of the revenue would be divided between Aerospace Technology Test and Demonstration, Media and PR, and Satellite Deployment – which happens to be very good news for XCOR, which is planning on a satellite-deployment capability for its Lynx Mark III.

THE FEDERAL AVIATION ADMINISTRATION'S OFFICE OF COMMERCIAL SPACE TRANSPORTATION (AST)

> "Many in the research community are hopeful to exploit the unique microgravity environment of suborbital flight with economical, routine access that enables expanded human research, atmospheric research, and microgravity biological and physical research. Space tourism proponents are optimistic that a safe, operational system will be developed to support their business ambitions. Yet there are significant technical, financial, and regulatory challenges to be overcome before these hopes can be realized. I encourage industry to work closely with the FAA, so that they will

be able to draft effective regulations in 2015, and diminish the chance that these regulations will stifle the industry."

The Honorable Steven M. Palazzo, Chairman Subcommittee on Space and Aeronautics Hearing on The Emerging Commercial Suborbital Reusable Launch Vehicle Markets

The Office of Commercial Space Transportation (AST) has been around longer than you might think, having been created in 1984. In those days, it came under the auspices of the Office of the Secretary of Transportation, which was in turn managed by the Department of Transportation. Then, in November 1995, the AST was moved to the FAA and was tasked with regulating the American commercial space transportation industry and making sure operators complied with the national security interests of the US. The AST is also responsible for encouraging and facilitating commercial space launches and strengthening the space transportation infrastructure of the US. And, if you happen to be in the business of building spacecraft, the AST is the one-stop shop you go to to apply for a launch and re-entry license – a process that is dealt with by the office's licensing and evaluation division. To begin with, operators such as XCOR and Virgin Galactic test their vehicles using an experimental launch permit issued by the AST. The administrative process by which an operator applies for launch and re-entry permits is described in the AST's regulations, section Title 14 CFR, Chapter III, Parts 415 (Launch License) and 431 (Launch and Re-entry of a Reusable Launch Vehicle). Let's take a look at Part 415 first.

Applying for a Launch License

Launching a rocket is very complex process – a fact reflected by the myriad considerations that the AST lists in the section that describes the process of applying for a license:

AST Part 415 Launch License[2]
Contents
Subpart A – General

§415.1 Scope.
§415.3 Types of launch licenses.
§415.5 Policy and safety approvals.
§415.7 Payload determination.
§415.8 Human space flight.
§415.9 Issuance of a launch license.
§415.11 Additional license terms and conditions.
§415.13 Transfer of a launch license.
§415.15 Rights not conferred by launch license.
§§415.16–415.20 [Reserved]

Subpart B – Policy Review and Approval

§415.21 General.
§415.23 Policy review.

---

[2] https://www.faa.gov/about/office_org/headquarters_offices/ast/licenses_permits/media/14cfr-401-417.pdf.

§415.25 Application requirements for policy review.
§415.27 Denial of policy approval.
§§415.28–415.30 [Reserved]

Subpart C – Safety Review and Approval for Launch From a Federal Launch Range

§415.31 General.
§415.33 Safety organization.
§415.35 Acceptable flight risk.
§415.37 Flight readiness and communications plan.
§415.39 Safety at end of launch.
§415.41 Accident investigation plan. §415.43 Denial of safety approval.
§§415.44–415.50 [Reserved]

Subpart D – Payload Review and Determination

§415.51 General.
§415.53 Payloads not subject to review.
§415.55 Classes of payloads.
§415.57 Payload review.
§415.59 Information requirements for payload review.
§415.61 Issuance of payload determination.
§415.63 Incorporation of payload determination in license application.
§§415.64–415.70 [Reserved]

Subpart E [Reserved]
Subpart F – Safety Review and Approval for Launch of an Expendable Launch Vehicle From a Non-Federal Launch Site

§§415.91–415.100 [Reserved]
§415.101 Scope and applicability.
§415.102 Definitions.
§415.103 General.
§415.105 Pre-application consultation.
§415.107 Safety review document.
§415.109 Launch description.
§415.111 Launch operator organization.
§415.113 Launch personnel certification program.
§415.115 Flight safety.
§415.117 Ground safety.
§415.119 Launch plans.
§415.121 Launch schedule.
§415.123 Computing systems and software.
§415.125 Unique safety policies, requirements and practices.
§415.127 Flight safety system design and operation data.
§415.129 Flight safety system test data.
§415.131 Flight safety system crew data.
§415.133 Safety at end of launch.

§415.135 Denial of safety approval.
§§415.136–415.200 [Reserved]

Subpart G – Environmental Review

§415.201 General.
§415.203 Environmental information.
§§415.204–415.400 [Reserved]

Appendix A to Part 415 – FAA/USSPACECOM Launch Notification Form
Appendix B to Part 415 – Safety Review Document Outline."

The above list gives you some idea of the many, *many* considerations involved in applying for a launch license. For those who are interested in the details of all these chapters and subparts, you can access the regulations at the AST website. But, to give you an insight into just one of these subparts, let's take a look at the requirements for human spaceflight (§415.8), which states:

"To obtain a launch license, an applicant proposing to conduct a launch with flight crew or a space flight participant on board must demonstrate compliance with §§460.5, 460.7, 460.11, 460.13, 460.15, 460.17, 460.51 and 460.53 of this subchapter."

To get started on demonstrating all those compliances the operator must check all the boxes against Subpart A (Launch and Re-entry with Crew) and Subpart B (Launch and Re-entry with a Space Flight Participant). For example, Subpart A is divided into Scope, Applicability, and Crew Qualifications and Training. Since most people will be interested in the crew aspect, we'll take a look at what the operator must do to satisfy the AST that its crew is qualified and trained to operate a SRV through its flight profile. One of the first requirements is that each crewmember must be able to fly the vehicle so the spacecraft doesn't harm the public, which is a simple enough requirement. But, after stating that condition, the requirements become more involved. The AST requires that crews be trained in normal and emergency scenarios, including abort and all manner of emergency scenarios. Crews must also demonstrate that they can deal with the myriad stresses of spaceflight, such as rapid-onset acceleration and deceleration, unusual attitudes, and microgravity. And, in case you have your eyes set on becoming a pilot of one of these SRVs and are dismissing that idea on the assumption that you need to be a retired NASA astronaut, that isn't the case. Far from it. The AST requires the pilot to be a certified FAA pilot with an instrument rating and also have the knowledge and skills to control the vehicle during a typical flight: that will probably require quite a few hours of flight experience together with plenty of time flying jets. In addition to accruing several thousand hours of flight time, it will probably help if you spend time in the simulator because the AST requires that pilots undergo mission-specific training. And, to cover itself in case things go pear-shaped, every pilot must sign a waiver of claims with the FAA in accordance with §460.19: Crew Waiver of Claims against the US Government.

Now, when it comes to Subpart B, the operator *has a lot* of informing to do. First, they must inform every passenger/prospective spaceflight participant about the many, *many* risks of launch and re-entry, and also provide details of the safety record of the SRV in question. So, if you happen to be a Virgin Galactic astronaut-in-waiting, Virgin Galactic is required by law to tell you about the crash of SpaceShipTwo. Operators must also go through the long list of known hazards and risks that passengers will be exposing themselves to during the flight as well as any hazards that may not be known. If you are a passenger waiting to fly, you will also be informed by your operator that it has its launch and re-entry license in order and also be given details of the number of catastrophic failures suffered by the vehicle you are about to board (hopefully this number will be zero!). The AST also requires that operators inform their customers of the number of passengers that have been killed and maimed during suborbital flights. Grim stuff. So let's move on to another operator requirement: spaceflight participant training. This subject has been open to some discussion partly because the AST requirement is far from specific. §460.51 (Spaceflight Participant Training) simply states:

> "An operator must train each space flight participant before flight on how to respond to emergency situations, including smoke, fire, loss of cabin pressure, and emergency exit."

We'll return to the subject of training in Chapter 8, so let's continue our look at the launch requirements. Once an operator has checked all the boxes for Subparts A through F, they then have to deal with the environmental issues, which includes everything from air quality and architecture to environmental justice and endangered animals. This can be a bit of a headache for the operator because it requires an environmental assessment to be conducted for the region of influence (ROI), which can include wetlands, floodplains, rivers, and farmland. It's a big area and in that area all sorts of endangered species may be present. For example, in the Mojave ROI, there exists the federally threatened *Gopherus agassizii*, aka desert tortoise, and the endangered ground squirrel, or *Xerospermophilus mahavensis*. The operator's job is ensure that launching and landing suborbital vehicles won't further endanger these animals, which means extensive surveys have to be conducted to prove these species haven't been detected within the ROI for at least several years.

Applying for a Re-entry License

So now that we've covered the requirements for launch, what does the AST have to say about re-entry? Well, re-entry is covered in Title 14, Chapter III, Part 431: Launch and Re-entry of a Reusable Launch Vehicle, Subparts B:

Contents

Subpart B – Policy Review and Approval for Launch and Re-entry of a Reusable Launch Vehicle

§431.21 General.
§431.23 Policy review.
§431.25 Application requirements for policy review.
§431.27 Denial of policy approval.
§§431.28–431.30 [Reserved]

Subpart C – Safety Review and Approval for Launch and Re-entry of a Reusable Launch Vehicle

§431.31 General.
§431.33 Safety organization.
§431.35 Acceptable reusable launch vehicle mission risk.

§431.37 Mission readiness.
§431.39 Mission rules, procedures, contingency plans, and checklists.
§431.41 Communications plan.
§431.43 Reusable launch vehicle mission operational requirements and restrictions.
§431.45 Mishap investigation plan and emergency response plan.
§431.47 Denial of safety approval.
§§431.48–431.50 [Reserved]

Subpart D – Payload Re-entry Review and Determination

§431.51 General.
§431.53 Classes of payloads.
§431.55 Payload re-entry review.
§431.57 Information requirements for payload re-entry review.
§431.59 Issuance of payload re-entry determination.
§431.61 Incorporation of payload re-entry determination in license application.
§§431.62–431.70 [Reserved]."

As you can see, there are lot of items to be checked off and a discussion of each of these is beyond the scope of this book, so let's focus on §431.35 (Acceptable Reusable Launch Vehicle Mission Risk) and §431.43 (Reusable Launch Vehicle Mission Operational Requirements and Restrictions), which happen to be particularly relevant to passengers. We'll begin with §431.35, an item that has been under increased scrutiny since the SpaceShipTwo tragedy. One of the problems of complying with this item is the definition of what is regarded as "acceptable risk." The AST simply states that "an applicant must demonstrate that the proposed mission does not exceed acceptable risk as defined in this subpart," but what does that mean? As far as the AST is concerned, risk is measured in the expected number of casualties and the precise number is 0.00003 casualties per mission. So, when you snuggle into your right seat on the Lynx, you can be reasonably assured that the chances of your not making it back to the runway in one piece are pretty remote. How will XCOR comply with this item? Well, it has to supply the AST with all the technical specifications of the Lynx together with details of radioactive materials, critical failure modes of systems, safety-critical events, and the consequences of the failure of any systems and/or materials. Of more relevance to the passenger are the risk and mitigation measures XCOR will employ to ensure passenger safety. For example, each passenger will wear a spacesuit (Figure 3.2), which will protect them in the event of a decompression event, and the flight profile of the vehicle is such that at no time will the vehicle's occupants be exposed to more than 4 Gs.

Another safety-driven item is §431.43 (Reusable Launch Vehicle Mission Operational Requirements and Restrictions), which requires that XCOR must conform to various procedures to ensure the Lynx is capable of operating outside of normal mission parameters in such events as an abort, for example. But complying with this item isn't simply ensuring that flight safety systems can be deployed manually because, in the event of an off-nominal incident, XCOR must also ensure that contingency abort locations are suitable for an impact. This means that factors such as debris dispersion and assessments of potential toxic release must be addressed. In addition to ensuring

3.2 PoSSUM scientist astronaut. Credit: Project PoSSUM

that a potential off-nominal event does not affect public safety, the AST also defines work and rest periods that require those operating the Lynx to have at least eight hours of downtime following a maximum 12-hour work shift. These work–rest periods allow for personnel to work no more than 14 consecutive days and no more than 60 hours in any seven-day period preceding a Lynx mission.

SPACEPORTS

Now that we have an insight into the process of applying for a launch and re-entry license, there is the business of the launch-site location. There are a number of launch sites for commercial vehicles in the US, nine of which are non-federal spaceports:

- California Spaceport at Vandenberg Air Force Base
- Cecil Field Spaceport, Jacksonville, Florida
- Kodiak Launch Complex on Kodiak Island, Alaska
- Mid-Atlantic Regional Spaceport at Wallops Flight Facility in Virginia
- Midland International Airport
- Mojave Air and Space Port in California
- Oklahoma Spaceport, Burns Flat, Oklahoma
- Spaceport Florida at Cape Canaveral Air Force Station
- Spaceport America, Las Cruces, New Mexico.

The application to operate a spaceport is a process that also falls under the purview of the AST, the details of which are noted in Title 14, Chapter III, Subchapter C, Part 420 (Licence to Operate a Launch Site). Since it is not the operator that applies for this license, Part 420 doesn't apply directly to XCOR, but passengers might be interested in knowing a little about their departure location. That location will be Midland International Airport, which received its spaceport license approval in September 2014, making it the first commercial service airport to be certified as a spaceport under Part 420. It is now known as the Midland International Air and Space Port (see Sidebar).

Midland International Air and Space Port at a Glance

- Classified as small-hub airport under Federal Aviation Regulation Part 139
- Certified as a spaceport by FAA under Part 420
- Supports American Eagle, Southwest Airlines, United Express
- 875 meters above sea level
- Four runways: 16R/34L: 2,896 × 46 meters

While it was Midland International Airport that applied for the spaceport license, a pivotal element in the application was XCOR, since they had planned on being the first tenant. Another tenant was Orbital Outfitters, a Hollywood-based company that manufactures spacesuits and space vehicle mock-ups (the company developed the Industrial Suborbital Spacesuit for use on the Lynx). We'll return to Orbital Outfitters in the training section in Chapter 6.

THE INTERNATIONAL TRAFFIC ON ARMS REGULATIONS (ITAR)

The ITAR has been a ball and chain around the neck of the American space industry for years. Originally developed to regulate military products and services, the ITAR now also cover many products that were initially developed for the military but have now become commercial products – navigation products, for example. At the heart of the ITAR is the USML (Table 3.2), which lists all sorts of products: weapons (Figure 3.3), military vehicles, flight control products … and a section labeled "satellites, launch vehicles and ground control equipment, including parts, technologies and software." This means that the Lynx is classified as a munition, which means XCOR, like all the other commercial spaceflight companies, must be very careful, because any violation could result in criminal liability and/or imprisonment. Needless to say, the ITAR/USML has caused much consternation in the commercial space industry partly because of the broad definition of Category XXI, which includes the following catch-all: *any other product, software, service or technical data with substantial military capability that was designed, developed, configured, adapted or modified for a military purpose*. It was this sort of language that caused Bigelow Aerospace's Mike Gold a major headache when dealing with a test stand for the company's Genesis inflatable habitat. In 2005, in preparation for the 2006 launch on top of a retired ballistic missile (the SS-20), Gold, Bigelow's corporate counsel, was in Russia

Table 3.2 The United States Munitions List (USML)

I – Firearms	XII – Fire Control, Range Finder, Optical and Guidance and Control Equipment
II – Artillery Projectors	XIII – Auxiliary Military Equipment
III – Ammunition	XIV – Toxicological Agents and Equipment and Radiological Equipment
IV – Launch Vehicles	XV – Spacecraft Systems and Associated Equipment
V – Explosives, Propellants, Incendiary Agents and Their Constituents	XVI – Nuclear Weapons Design and Related Equipment
VI – Vessels of War and Special Naval Equipment	XVII – Classified Articles, Technical Data and Defense Services Not Otherwise Enumerated
VII – Tanks and Military Vehicles	XVIII – Reserved
VIII – Aircraft and Associated Equipment	XIX – Reserved
IX – Military Training Equipment	XX – Submersible Vessels, Oceanographic and Associated Equipment
X – Protective Personnel Equipment	XXI – Miscellaneous Articles
XI – Military Electronics	

3.3 SM-3 missile. Credit: USN

with the Genesis test stand, a metal sheet with four legs (he had gotten that far thanks to a Technical Assistance Agreement). Turn it over and you had a coffee table. But this particular coffee table, since it was designated as a test stand, had a dual purpose, which meant Gold had to have two guards watching it 24/7 – just in case the Chinese got hold of the militarily sensitive technology and repurposed it by … serving coffee on it perhaps?

But the ITAR/USML issue extends beyond materiel, which Virgin Galactic discovered to their cost. In 2012, Virgin Galactic featured in the commercial space blogs with the news that the company was not allowed to fly Chinese nationals. Why? The ITAR.

> "Virgin Galactic adheres to both the spirit and the letter of US export controls and has for now chosen not to accept deposits from countries subject to US export and other regulatory restrictions."

This case was related to Part 126.1 of the ITAR which prohibits the export of technologies under its control to selected nations, one of which is China. This had nothing to do with Virgin Galactic flying SpaceShipTwo from China – something that would most definitely be prohibited under the ITAR. The reason for not being able to sell tickets to the Chinese was that such an act would fall foul of the export regulations, since the ticket would be classified as a "related item" under the list of prohibited items on the USML. And, since XCOR is in the same business as Virgin Galactic, it is unlikely you will be seeing any Chinese passengers climb on board the Lynx anytime soon. While there are some who argue that the restrictive language of the ITAR/USML is not helping the industry, those restrictions are in some way offset by the government's regulatory regime for suborbital flight, which has helped the industry. Perhaps, but the ITAR is still a thorn in the side of suborbital commercial spaceflight. There was some progress in 2013 when Virgin Galactic's flight operations were removed from the control of the ITAR. This meant that the company could now fly non-US citizens without having to jump through the hoops necessary to get an export license (under the old rules, even Sir Richard Branson wouldn't have been allowed to fly on his own spacecraft!). This also appeared to be good news for XCOR because it would make international operations such as those planned for Curaçao much easier, but the news wasn't quite as rosy as it first appeared. That's because the exemption that Virgin Galactic received stated that as long as the company's hardware was built in the US the government would have authority over anything the company sent abroad. It was for this reason that XCOR developed the wet lease concept on which the company's agreement with Space Experience Curaçao is based. This dictates that the Dutch will pay American (XCOR) personnel to operate the Lynx. If XCOR were to do otherwise, they would end up being traffickers in controlled goods and that wouldn't be good for business.

Since XCOR's wet lease agreement, the US space industry has continued to lobby the government to get the ITAR/USML regulations amended and, in December 2012, that lobbying paid off – at least for the satellite industry. Congress struck out the legalese that placed satellites on the USML, although the prohibitions on export remained. But, in its new draft, Congress added "man-rated suborbital, orbital, lunar and interplanetary spacecraft" to Category XV of the USML. It was a case of two steps forward one step back, with the result that the lobbying continued unabated. In April 2015, a commercial space advisory group submitted a proposal to remove manned commercial spacecraft from the USML at a meeting of the FAA's Commercial Space Transportation Advisory Committee.

> "The U.S. space industry will benefit from placing human spaceflight systems under the auspices of the Export Administration Regulations."
>
> *Mark Sundahl, Chairman of COMSTAC's International Space Policy Working Group*

After much deliberation among members that included Gold (COMSTAC Chairman) and Jeff Greason, the committee agreed to recommend that manned commercial suborbital vehicles should be removed from the ITAR's jurisdiction. Whether Congress approves remains to be seen!

4

Next-Generation Spacecraft

Credit: XCOR

Imagine a spacecraft that can take off from a runway, rocket to suborbital altitudes, and then glide gracefully back to the same runway. It's a concept that seemed out of reach even 10 years ago but, thanks to the dedicated vision of XCOR and its band of hard-working and gifted engineers, that dream is taking shape in the form of the Lynx. No strap-on boosters or expendable tanks for this puppy. The Lynx, a thoroughbred fully reusable launch vehicle (RLV), can make it to 100 kilometers of altitude under its own steam – a

E. Seedhouse, *XCOR, Developing the Next Generation Spaceplane*,
Springer Praxis Books, DOI 10.1007/978-3-319-26112-6_4

capability that gives it an advantage over other suborbital vehicles such as Virgin Galactic's SpaceShipTwo, which relies on an air-launch system to lift the vehicle to its 15,000-meter release altitude. And then there's the low operating costs and short turnaround times that enable this new breed of suborbital spaceship to launch up to four times a day. And, whereas you have to drive to the middle of nowhere to get to Spaceport America, which is the operating ground for Virgin Galactic, to take your flight on board the Lynx you can fly commercial to Midland International. Powered by four rocket engines burning liquid oxygen and kerosene, the Lynx will reach Mach 2.9 just three minutes into the flight. The engines will then be turned off and momentum will do the rest, with the Lynx coasting up to 100,000 meters of altitude, where the sole passenger and pilot will experience a little over four minutes of weightlessness. All for just US$150,000.

LYNX VARIANTS

There will be three Lynx variants (see Sidebar), the first of which – the Mark I – is nearing completion as this book is being written. The Mark I will serve as a prototype spacecraft and test vehicle for the Mark II (Figure 4.1) and Mark III. The test flight program for the Mark I will feature up to 80 flights and take between 6 and 18 months, during which time the equipment and systems will be continuously tested and evaluated until the vehicle is capable of demonstrating the complete flight profile. That flight profile will not feature a trip into space, since the maximum altitude of the Lynx Mark I will be 61,000 meters, which is about two-thirds of the way to space. But those test flights will be invaluable, since they will allow XCOR's test pilots to gain priceless experience that can be transferred to the Mark II, which will be the suborbital version, capable of flying to altitudes of at least 100,000 meters. XCOR's plan is to develop the Mark II in parallel with the flight testing of the Mark I and, once the Mark II is up and running, XCOR plan to develop the Mark III

4.1 Lynx Mark II. Credit: XCOR

4.2 Lynx Mark III with the Atsa Observatory. "Atsa" is the Navajo word for "eagle," incidentally. Credit: XCOR

(Figure 4.2), a beefed-up version of the Mark II that will feature an enhanced payload capability thanks to a dorsal pod capable of carrying up to 650 kilograms of payload.

The Lynx, Mark by Mark

Mark I. Prototype. Due to commence its flight testing program at the end of 2016, this variant will be used to test propulsion, life-support systems, spacecraft structure, re-entry heating, and the aeroshell. After the vehicle is licensed by the Federal Aviation Administration (FAA), it will be brought into commercial service.

Payload time in microgravity: 63 seconds below 0.001 g_o

Mark II. Production vehicle. This version is intended to serve the suborbital tourism, payload, and science markets. It will use the same avionics and propulsion system as the Mark I but will weigh less thanks in part to a special lightweight liquid-oxygen tank and other lightweight proprietary innovations.

Payload time in microgravity: 133 seconds below 0.001 g_o

28 seconds below $1 \times 10^{-6} g_o$

Mark III. Derivative vehicle. This will be a modified Mark II with the added feature of an external dorsal pod (the "payload integrator") capable of launching a small satellite into low Earth orbit (LEO) or carrying a payload experiment. Other modifications will include enhanced landing gear, better aerodynamics, structural enhancements, and a propulsion package that will pack a more powerful punch than the one fitted on the Mark II.

Payload time in microgravity: 133 seconds below 0.001 g_o
28 seconds below 1×10^{-6} g_o

LYNX PERFORMANCE

The Lynx will carry just one pilot and one passenger, each of whom will be required to wear spacesuits and who will be strapped in for the flight's duration. This is in marked contrast with SpaceShipTwo, which will be capable of carrying up to six passengers and two crew. SpaceShipTwo has the added bonus of allowing the passengers to float around during their four minutes of weightlessness. Having said that, the cost of a ticket on SpaceShipTwo is US$250,000 and that amount doesn't guarantee you a trip to space. If you happen to be a Virgin Galactic passenger, you may want to read the small print on your ticket that states that the company guarantees it will fly you to an altitude of at least 50 miles. Well, 50 miles, or 80 kilometers, is not the internationally recognized altitude of space, which starts at 100 kilometers. Virgin Galactic CEO George Whitesides cleared up the issue in May 2014 by issuing the following statement to Gizmodo:

> "NASA and the US Air Force have a long tradition of celebrating everything above 50 miles (~80km) as spaceflight, and we look forward to joining those ranks soon as we push onward and upward. We are still targeting 100km. As we have always noted, we will have to prove our numerical predictions via test flights as we continue through the latter phase of the test program. Like cars, planes, and every other type of vehicle designed by humans, we expect our vehicle design and performance to evolve and improve over time. When SpaceShipTwo reaches space for the first time – which we expect will happen just a few short months from now – it will become one a very small number of vehicles to have ever done so, enabling us to commence services as the world's first commercial spaceline; our current timetable has Richard's flight taking place around the end of the year."

Of course, Virgin Galactic didn't reach space in 2014 because SpaceShipTwo crashed later that year, killing one of the pilots and seriously injuring another. And the problem with the altitude? Seems part of the problem was the rubber-based fuel because, shortly after George Whitesides's statement, Virgin Galactic announced it was switching from the rubber-based solid fuel (HTPB) to a plastic-based fuel (polyamide). Still, the fuel switch still didn't alter the small print on the ticket, although Virgin Galactic were at pains to point out that any of their passengers flying higher than 80 kilometers of altitude would be awarded astronaut wings: Virgin Galactic astronaut wings, not FAA commercial astronaut wings, though. But let's get back to the performance of the Lynx variants.

Lynx Mark I

We'll start with the flight profile which begins like any aircraft flight profile with an acceleration along a runway. Take-off speed is 190 knots. That's a lot faster than your average 737, which takes off at around 130 knots. But perhaps a better comparison is to compare the Lynx with a corporate jet that is about the same size – the Embraer Phenom 100, for example. As you can see in Table 4.1, the Phenom (Figure 4.3) is about the same weight as the Lynx but that's where the similarities end.

As you can imagine, the speed at which the Lynx operates and the rate of climb (Figure 4.4 and Table 4.2) all add up to some G-loading (Figures 4.5a and 4.5b), which lasts until engine cut-off at 2 minutes and 38 seconds into the flight, so it's worth taking a look at how much punishment will be inflicted on the pilot and passenger.

We'll cover the business of acceleration training in Chapter 6, but there may be some reading this who are wondering what is meant by the *Z*-axis and the *X*-axis in Figure 4.5a and 4.5b, so here goes. We'll start by looking at the neat diagram (Figure 4.6) provided by

Table 4.1 The Embraer versus the Lynx

Performance	*Embraer Phenom*	*Lynx*
Length	12.8 m	8.51 m
Span	12.3 m	7.3 m
Height	4.3 m	2.2 m
Gross weight	4,770 kg	4,850 kg
Maximum operating speed	463 knots or Mach 0.70	Mach 2
Take-off distance	952 m	365 m
Service ceiling	12,497 m	61,000 m (Mark I)
Fuel type	Jet A, Jet A-1, or JP-8	Liquid oxygen/kerosene

4.3 The Embraer Phenom jet. Credit: Magnus Manske

Table 4.2 Lynx Mark I flight profile[1]

Elapsed time (sec)	*Milestones*
0	Engine start
13	Take-off
158	Engine shutdown. Vehicle coast phase
194	Low acceleration phase ~0.001 g_0
234	Apogee at 61,000 m. Beginning of free fall
257	Acceleration exceeds 0.001 g_0
305	Onset of pull-out of 1 g
1,500	Touchdown

[1]Adapted from XCOR's *Lynx Payload User's Guide Version 3b*, 24 July 2012.

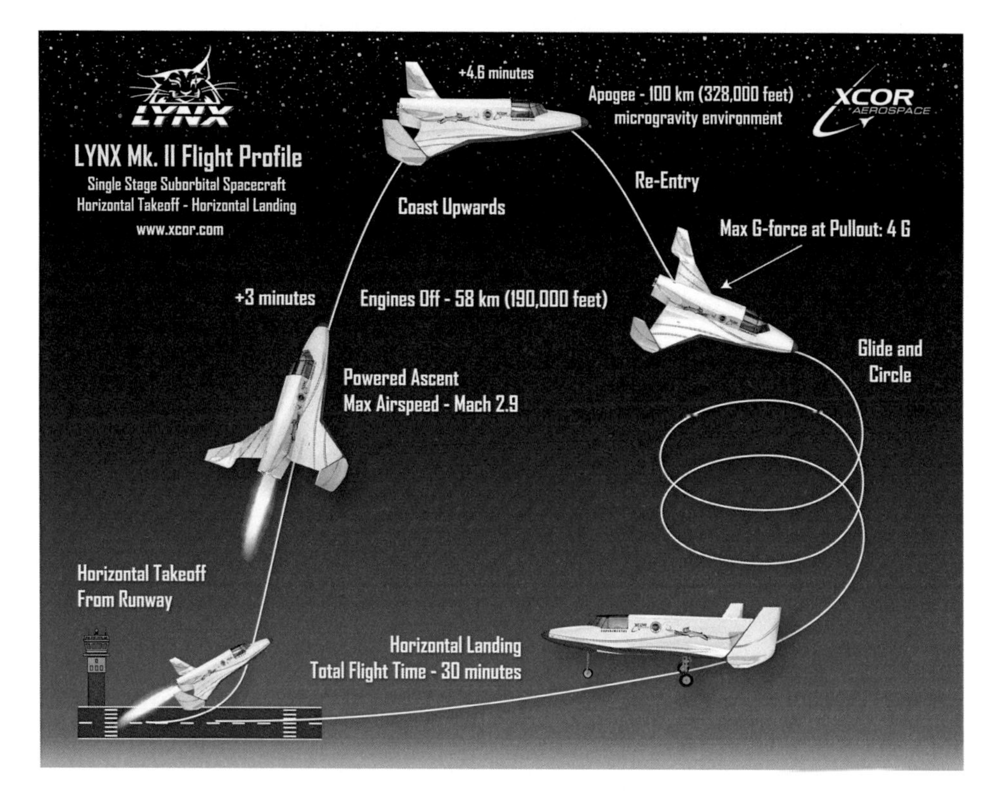

4.4 Lynx Mark I profile. Credit: XCOR

NASA which sums up much of what I have to say in this section. As you can see, G-forces act on pilots and passengers in three axes – *X*, *Y*, and *Z* – each of which has a positive and a negative direction. When you are standing, the force of gravity acts on the longitudinal axis parallel to your back but, when you are in a spacecraft, the force of gravity, or G, acts in different axes, depending on whether the vehicle is yawing (G*z*), rolling (G*x*), or

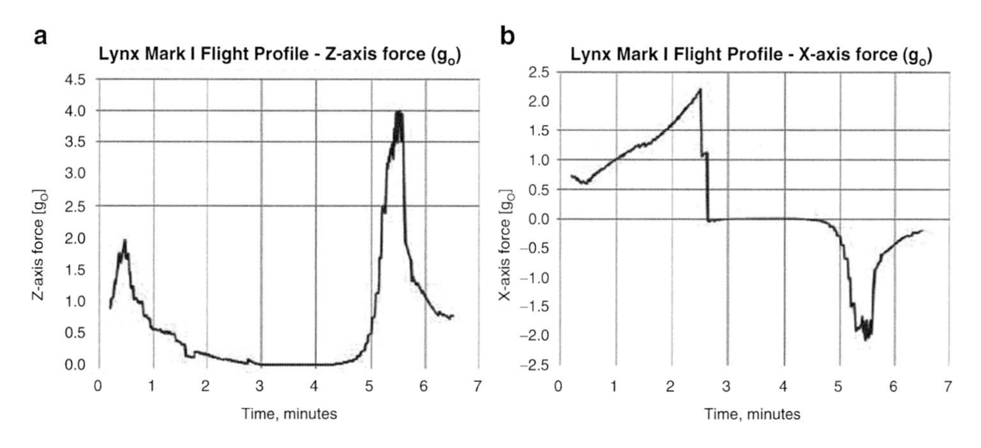

4.5 Lynx Mark I flight profile showing the (**a**) Z-axis and (**b**) X-axis. Credit: XCOR

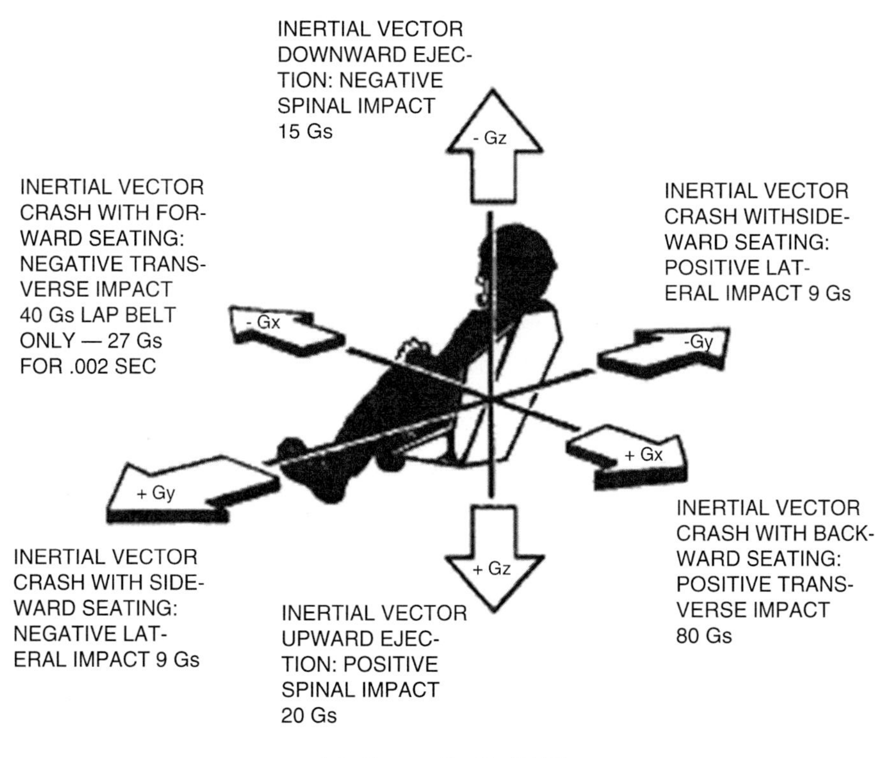

4.6 G-axes. Credit: NASA

pitching (G*y*). When you are sitting in the Lynx as it is accelerating upward, you will feel acceleration in the G*x* axis. And, as you can see in Figure 4.6, the maximum *X*-axis force is about 2.2 Gs, which means that, if you weigh 70 kilograms and you are pulling 2.2 Gs, you will actually weigh 154 kilograms.

Lynx Mark II

The Mark II (Figure 4.7 and Table 4.3) flies the same flight profile (Figure 4.8) as the Mark I with just a few differences. First, the Mark II will be accelerating for longer than the Mark I and, secondly, this vehicle reaches suborbital altitude, which also means it takes longer to get back down to Earth. As you can see in Figures 4.9a and 4.9b, the G-loading

Table 4.3 Lynx Mark II flight profile[1]

Elapsed time (min/sec)	*Milestones*
0	Engines start
00:13	Take-off
03:01	Engine shutdown. Vehicle coast phase
03:34	Low acceleration phase ~0.001 g_o
04:32	Begin microgravity period below 10^{-6} g_o
04:46	Apogee at 61,000 m. Beginning of free fall
05:00	Acceleration exceeds 0.001 g_o
05:47	Onset of pull-out of 1 g
27:02	Touchdown

[1]Adapted from XCOR's *Lynx Payload User's Guide Version 3b*, 24 July 2012.

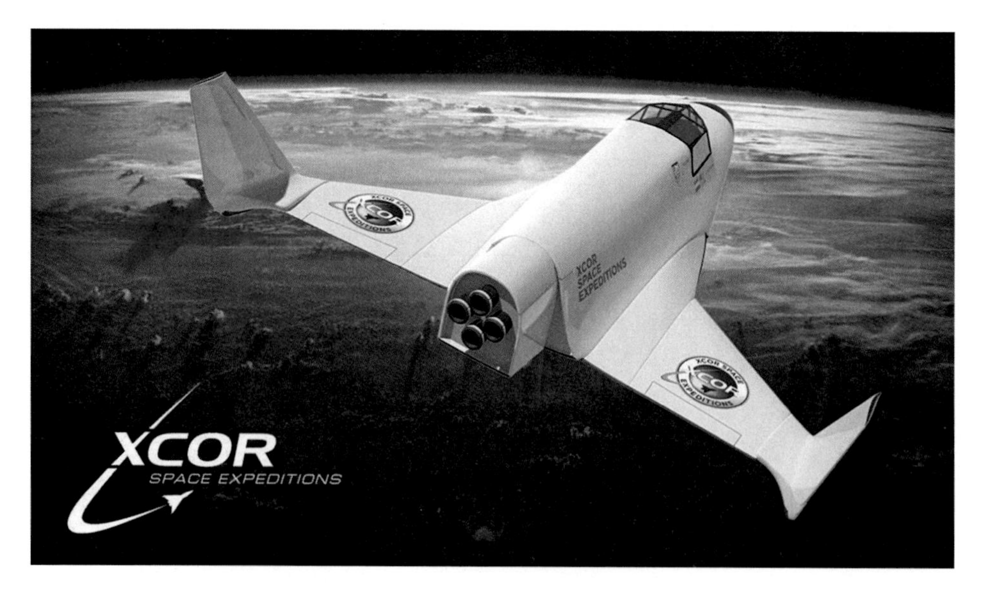

4.7 Lynx Mark II. Credit: XCOR

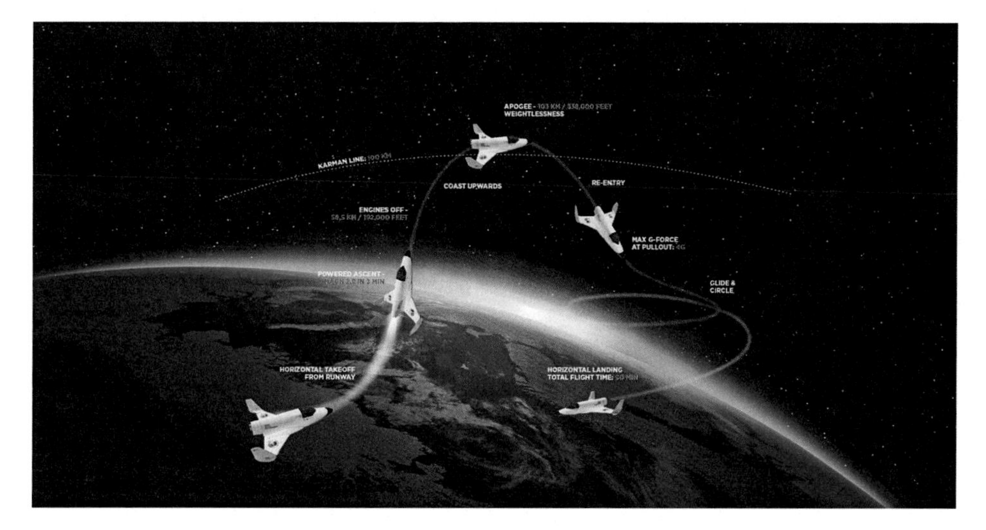

4.8 Lynx Mark II flight profile. Credit: XCOR

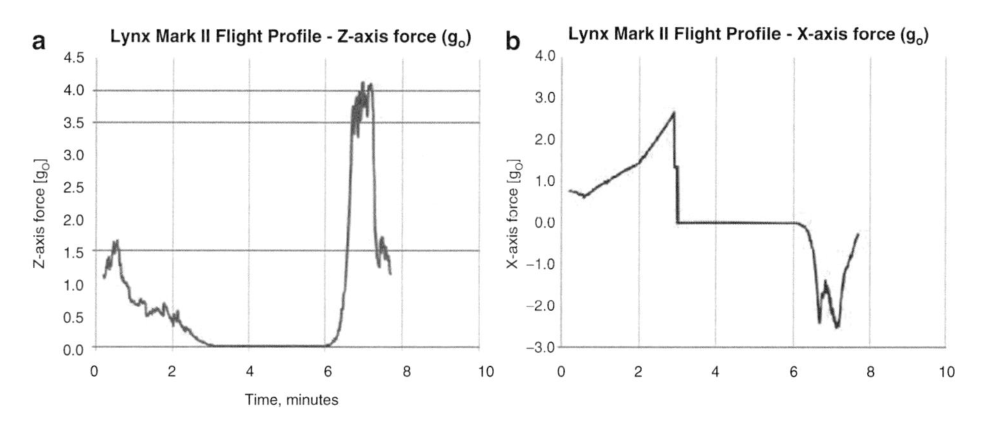

4.9 Lynx Mark II flight profile showing the (**a**) *Z*-axis and (**b**) *X*-axis. Credit: XCOR

isn't that much different, the main difference being that the pilot and passenger will be under G for a longer period on the way up and on the way down, so it will pay to practice that anti-G straining maneuver (AGSM) that will be described in Chapter 6.

Lynx Mark III

This vehicle flies practically an identical profile to the Mark II with the exception that this vehicle is fitted with that dorsal pod so it can release a satellite or expose a payload to a vacuum. As with the Mark I and Mark II, control of the Mark III will be under the direction of the pilot, who will manually control the vehicle using rudder controls, trim flaps, and drag brakes. When the air density is too low to use these control surfaces, the pilot will control pitch, yaw, and roll using the Lynx's reaction control system (RCS), which will enable the pilot to orient the Lynx in just about any direction and stabilize the vehicle in just 10 seconds.

PAYLOADS

As you can see in Figures 4.10a and 4.10b, the Lynx can be configured for a variety of internal and external payloads (Appendix III). While the Mark I and Mark II can carry up to 280 kilograms, the Mark III can loft 650 kilograms to 100 kilometers. XCOR flies primary and secondary payloads: the primary payloads drive the flight trajectory and mission objectives, while the secondary payload, which has no bearing on trajectory or mission objectives, is loaded together with the primary payload.

If you're interested in flying a payload, you have the option of locating it inside the cabin or outside the cabin. But no matter where you want to locate your payload (Table 4.4), you must make sure that whatever you want to fly complies with XCOR's integration standards for size, strength, containment, and vehicle safety. These integration standards also include the positioning of tethers, cable routing, and any other items that may not be carried as part of a regular Lynx flight. If Payload B (which sits next to the pilot) happens to be a spaceflight participant conducting an experiment, then it is worth testing how much room there will be for that spaceflight participant to manipulate whatever equipment needs manipulating. That's because, when you're wearing a spacesuit, the confines of the Lynx cockpit are rather cozy (Figure 4.11), as Project PoSSUM candidates discovered when they beta-tested their experiment in Embry-Riddle's suborbital Lynx trainer.

Payload A

This payload (Figure 4.12) is attached to a track seat just behind the pilot's seat. Its shape approximates to a triangle with the top cut off. This payload can carry up to 20 kilograms and the most popular use for this area will most likely be science experiments carried in the Cub Carrier. The Lynx Cub Payload Carrier, to give it its full name, was designed in 2013 to carry small experiments as secondary payloads: anything from fluid physics to materials processing to life sciences. The initiative, which was developed by the United States Rocket Academy and the Space Engineering Research Center, with support from XCOR, allows very easy integration of experiments and very simple interfaces. Best of all, the Cub Carrier, which has room for ten 10 × 10 × 10 centimeter boxes, is very affordable: just US$3,000 for each box.

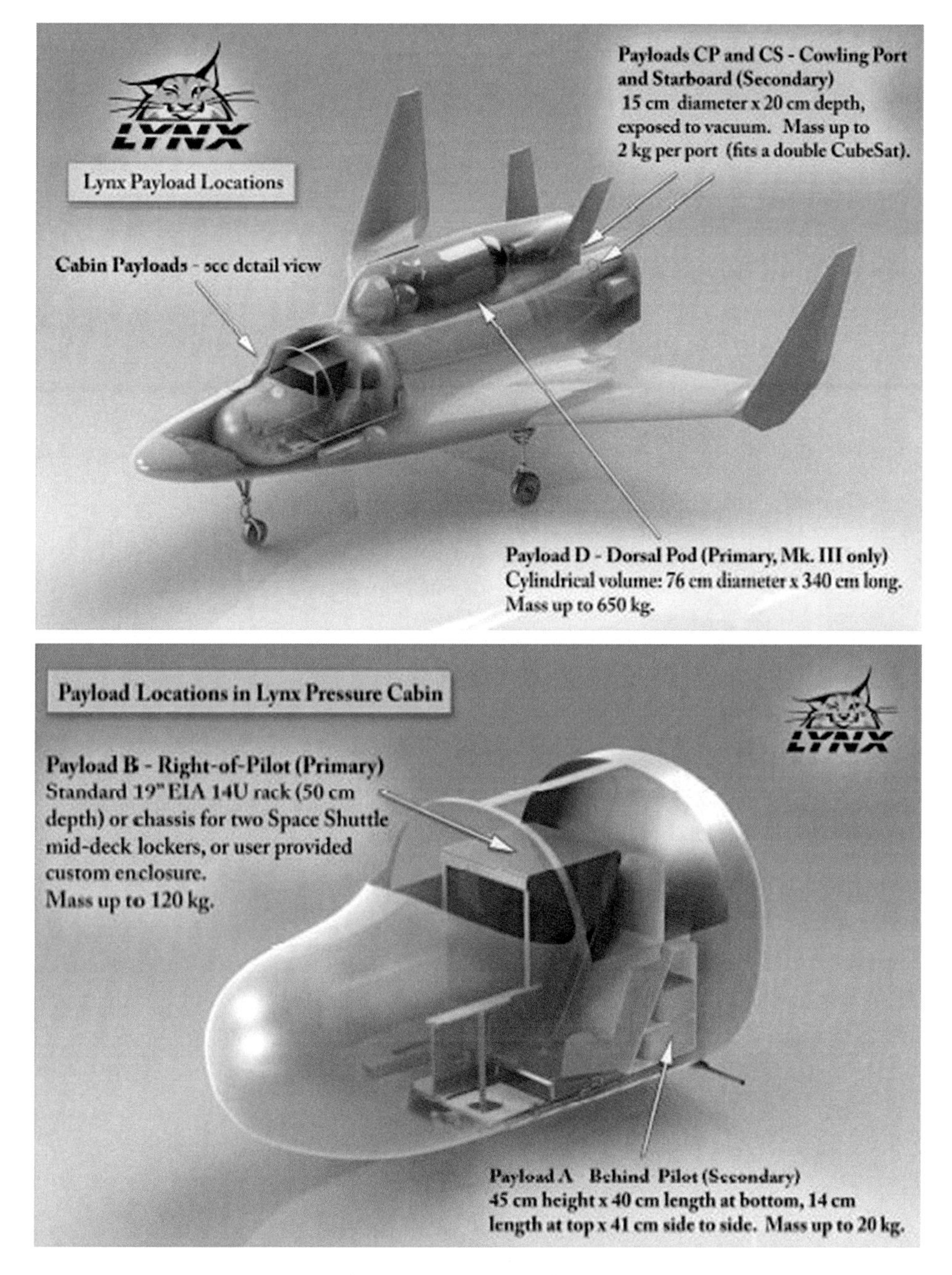
LYNX
Lynx Payload Locations
Payloads CP and CS - Cowling Port and Starboard (Secondary)
15 cm diameter x 20 cm depth, exposed to vacuum. Mass up to 2 kg per port (fits a double CubeSat).
Cabin Payloads - see detail view
Payload D - Dorsal Pod (Primary, Mk. III only)
Cylindrical volume: 76 cm diameter x 340 cm long.
Mass up to 650 kg.
Payload Locations in Lynx Pressure Cabin
LYNX
Payload B - Right-of-Pilot (Primary)
Standard 19" EIA 14U rack (50 cm depth) or chassis for two Space Shuttle mid-deck lockers, or user provided custom enclosure.
Mass up to 120 kg.
Payload A Behind Pilot (Secondary)
45 cm height x 40 cm length at bottom, 14 cm length at top x 41 cm side to side. Mass up to 20 kg.

4.10 Lynx payload. Credit: XCOR

Table 4.4 Payload integration locations[1]

Location	*In cabin/external*	*Lynx Marks I and II*	*Lynx Mark III*	*Limits to other payloads*
Payload A	In cabin	20 kg Height: 45 cm Width: 41 cm Top depth: 14 cm Bottom depth: 40 cm	20 kg Height: 45 cm Width: 41 cm Top depth: 14 cm Bottom depth: 40 cm	No limitations on other payloads
Payload B	In cabin	120 kg Standard EIA[a] rack (41-cm depth) or chassis for two Shuttle middeck lockers	120 kg Standard EIA[a] rack (41-cm depth) or chassis for two Shuttle middeck lockers	Cannot carry a spaceflight participant in this configuration
Payloads CP and CS[b]	External	2 kg each Diameter: 15 cm Depth: 20 cm	2 kg each Diameter: 15 cm Depth: 20 cm	No limitations on other payloads
Payload D	External	Not applicable	650 kg at second-stage ignition Cylindrical volume: 76 cm Length: 340 cm	Maximum mass precludes other payloads

[1]Adapted from XCOR's *Lynx Payload User's Guide Version 3b*, 24 July 2012.
[a]Electronic Industries Alliance EIA-310-D, Cabinets, Racks, Panels and Associated Equipment.
[b]Cowling Port, Cowling Starboard.

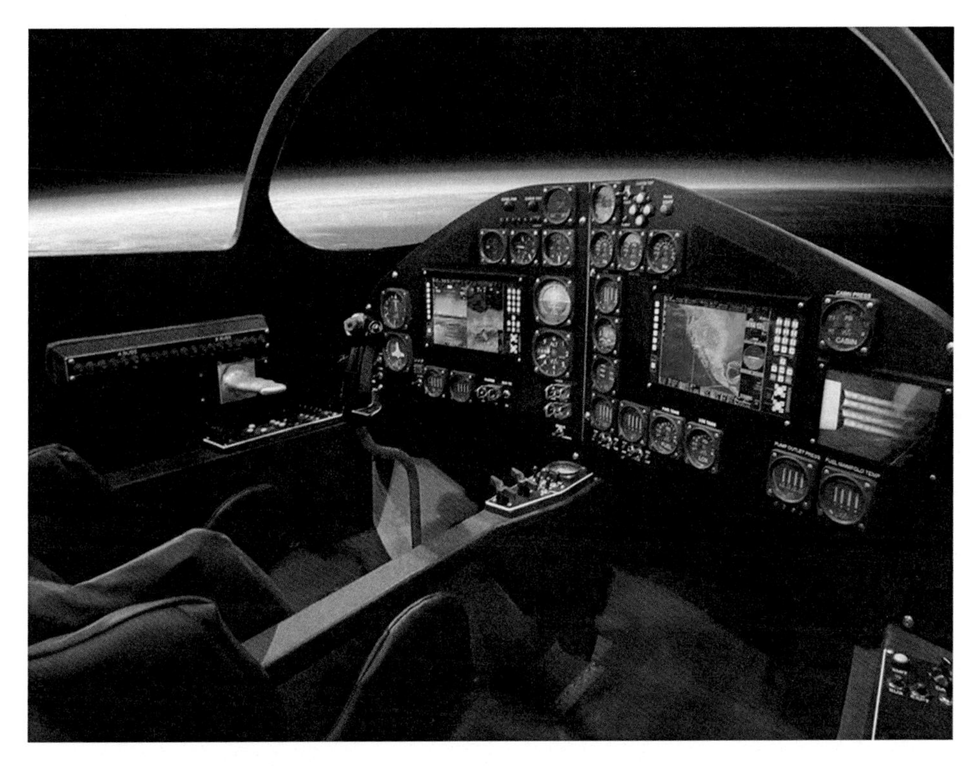

4.11 Lynx cockpit. Credit: XCOR

4.12 Lynx cockpit close-up. Credit: XCOR

The Cub Carrier also happens to be a very versatile system, comprising off-the-shelf hardware that can accommodate just about any conceivable payload as long as it fits inside one of those cubes and weighs less than one kilogram. Want to film your payload? No problem: the Cub Cam will take care of it. Need electrical power during the flight? The Cub Carrier provides 5 or 12 V, which is configurable prior to flight. And, if you would like to track your payload in real time, then wireless or wired control is available. Alternatively, if you just want to kick back and let the experiment run without any inputs, an autonomous control option is also available. Once flights begin, XCOR will gather baseline data for those designing experiments and these data will provide information about the acoustic and pressure environment together with temperature and radiation limits encountered during a typical flight.

Payload B

In most flights, Payload B will be a spaceflight participant or scientist. As you can see in Figure 4.13, the passenger will be seated in the right seat and they will be wearing a spacesuit which will most likely have been manufactured by either Final Frontier Design (FFD) or Orbital Outfitters. If you happen to have a compact frame such as Jonna, pictured in

4.13 Jonna, a Project PoSSUM-trained scientist-astronaut in Embry-Riddle Aeronautical University's Suborbital Spaceflight Simulator, February 2015. Credit: Jason Reimuller

Figure 4.13, then you will have some room to maneuver but, if you happen to be a big guy, then movement will be minimal, so bear that in mind when you're planning your experiment. If you happen to be flying in the right seat conducting an experiment, you will need to be familiar with the cabin environment, specifically the acoustic levels, acceleration/deceleration forces, temperature (this will be around 20°C), air pressure (10.5 psi or equivalent to 2,750 meters), air composition, humidity (between 20 and 50%), radiation (negligible), contamination, and vibration. We'll talk about these in more detail in Chapter 6.

In the event that the right seat is occupied by a payload, the right seat will be removed and an experiment rack will be installed by locking it into place using seat tracks. The maximum mass for Payload B is 120 kilograms and the payload can be contained either within a custom unit or within a standard payload container. The standard payload container will fit two Space Shuttle middeck lockers and the details of how payloads can be integrated in these lockers are a subject covered in Chapter 6.

Payload Cowling Port and Cowling Starboard

The Lynx can carry two cowling payloads that can weigh up to two kilograms each and measure no more than 10 × 10 × 20 centimeters. The payloads, which must fit into a cylindrical volume (Figure 4.14), will be loaded into the Lynx's cowling ports just before the pilot conducts his final briefing. If you are interested in flying a CP or CS payload, then bear in mind that no environmental controls are available for these payloads.

Payload D: Dorsal Pod

These payloads (Table 4.5) will be mounted inside the Lynx's (Mark III) dorsal pod fairing, as shown in Figure 4.14. In common with the CS and CP payloads, the dorsal pod will have no environmental controls, although power will be available in case the researcher needs to heat or cool the payload. As you can see in Figure 4.14, the dorsal pod has a cylindrical collar and the *Y* and *Z* dimensions are tapered, so these size restrictions have to be borne in mind when configuring the payload. If your payload happens to be a satellite that will be deployed, then you need to be familiar with the mode of loading and release. The pod features two hinged sections, one being the nose of the pod, which opens to the front, and the other the back of the pod, which is hinged to open to the rear. To load your payload, you simply slide in from the front of the spacecraft.

If you happen to be launching your payload (which will cost you around US$500,000, incidentally), the mode of release begins with the opening of the forward fairing cover followed by the opening of the rear cover to permit either a rocket-powered or spring launch (launched payloads cannot be retrieved by the Lynx, incidentally). Payloads that remain inside the pod for the flight's duration will be held down by clamps.

BUILDING THE LYNX

Now let's dial the clock back to February 2014. If you had stepped into the XCOR hangar that month, you wouldn't have seen anything resembling a spaceship, but the pieces were there. A hydrogen stand, the fuselage, the liquid-oxygen tank, the strakes (we'll get to these

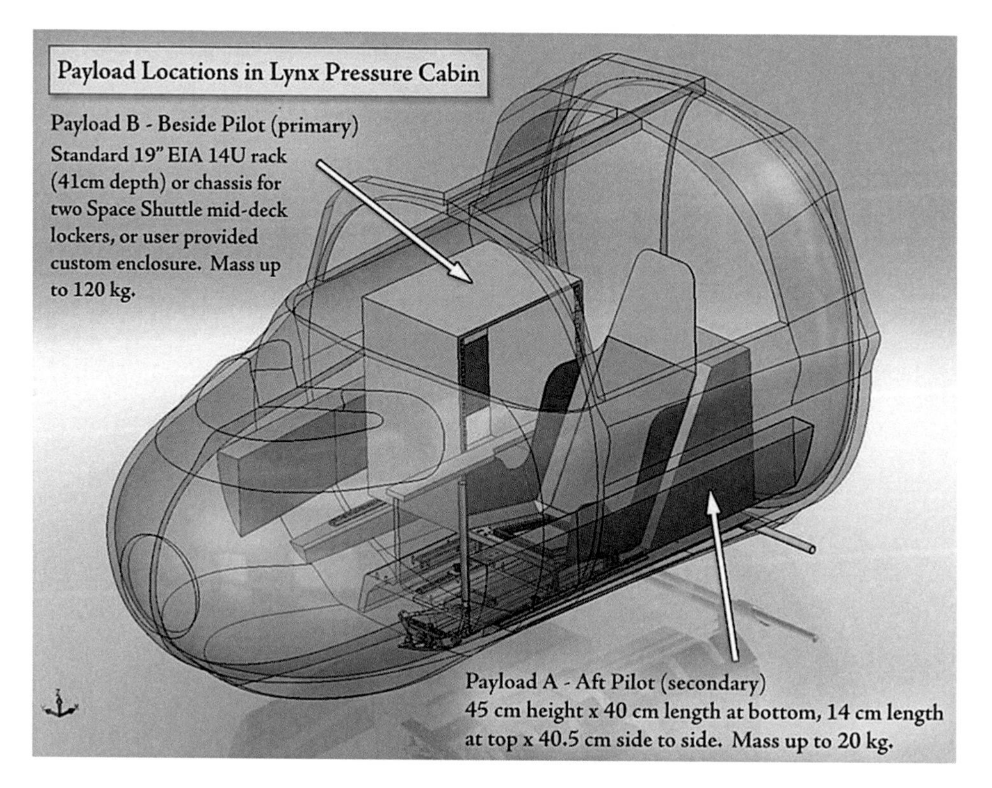
Payload Locations in Lynx Pressure Cabin
Payload B - Beside Pilot (primary)
Standard 19" EIA 14U rack (41cm depth) or chassis for two Space Shuttle mid-deck lockers, or user provided custom enclosure. Mass up to 120 kg.
Payload A - Aft Pilot (secondary)
45 cm height x 40 cm length at bottom, 14 cm length at top x 40.5 cm side to side. Mass up to 20 kg.

4.14 Lynx payload locations. Credit: XCOR

Table 4.5 Dorsal pod specifications

Payload D	*Dorsal pod*
Volume under dorsal pod	Diameter: 76 cm
Length: 340 cm	
Payload launch/separation options	Spring, gas-, or rocket-powered stage
Maximum weight of payload and stage	650 kg
Nominal altitude	400 km

shortly), and the landing gear: if you take a look at Figure 4.15, you can see where all these bits and pieces fit together. One element that was missing was the cockpit, but it was on its way, which meant that XCOR could finally look forward to final assembly. But, before we discuss the assembly process – and those strakes – it's worth explaining why the process of designing, developing, and ultimately building a spaceship takes so much time. By now, most of you reading this will be familiar with the predictions of rocket companies declaring launch dates, only to see those dates slide to the right again and again. Sometimes these companies are their own worst enemy because building spacecraft takes time, which is why

4.15 The Lynx Mark I fuselage, rear view. Credit: XCOR

you have to applaud XCOR for taking the "we'll fly when we're good and ready" approach. No glitzy pronouncements of when revenue flights will start from Jeff Greason and his talented team of engineers. Remember, we're talking about building a prototype spacecraft, not a do-it-yourself aircraft build. And building a new vehicle requires an awful lot of groundwork that demands extensive research and development of test articles which invariably have to be redeveloped and retested before a final item can be fabricated. It's an extremely time-consuming process. And at the core of that process is the fuselage (Figure 4.16), which has to be rugged and strong enough to stand up to not only the very harsh environment of space, but also the very demanding flight to and from that destination.

The fuselage must also be strong because all the other major structural components attach to it. For example, the truss structure that houses the propulsion system, the cockpit, the wings, the strakes, and the nose structure are all attached to the fuselage. But it's not just a case of slotting these components together. In the case of a spacecraft, form must follow function, which means each component has its own functional requirements, but at the end of the day these highly engineered structures must fit together seamlessly so that the end result is a high-performance vehicle capable of safely transporting its crew and payload to and from space. And to achieve that goal is anything but easy. So this is what XCOR's engineers were working towards in February 2014. But, before final assembly could commence, XCOR's engineers had to get on with the job of piecing together all the structural subassemblies while simultaneously debugging the propulsion system. With that done, the next item on the agenda was working around those strakes. The strakes comprise the main structure that attaches the wings to the Lynx. The Lynx has one strake

4.16 The Lynx Mark I fuselage, front view. Credit: XCOR

on the port side and one on the starboard side, and each strake contains four kerosene fuel tanks, the landing gear, and a pair of reaction control thrusters. Like every element of the Lynx, the strakes had been subjected to all sorts of tests before the final articles were fabricated. For example, slosh tests, which were conducted in 2013, had to be performed to help engineers figure out where baffles needed to be installed to control the sloshing of fuel against the strake walls.

In March 2014, XCOR's engineers were configuring the plumbing and electrical work that passed through the strakes and moving towards the installation of the landing gear, which was hung from titanium beams that were in turn mounted inside the strakes. With the landing gear installed, engineers then had to perform multiple retraction and deployment cycles to make sure the system worked without issue. It sounds straightforward but nothing is clear-cut in the "building a spacecraft from scratch" game. One of the biggest challenges of building the Lynx was working with carbon fiber and epoxy, which meant that, in the case of the four fuel tanks, the tanks and the attachment points all had to be built at the same time. And then there was the challenge of actually building those parts … or baking them in this case. The engineers would close out the parts and put them in the oven, praying that the components didn't move during the process. And those strakes? Those took almost three years to design before engineers had the fabricated articles in front of them. Why carbon fiber? In short, carbon fiber is light and very stiff, which happen to be ideal qualities when designing a spacecraft. The problem is that the bonding process is challenging and it can be a real pain laying the cloth in a way that ensures the flight loads are transferred optimally (which is why XCOR's engineers used mechanical fasteners to reinforce areas of high load such as joints).

The business of working towards final assembly progressed rapidly throughout 2014. By the end of the year, engineers had bonded the cockpit (see sidebar) to the fuselage and were preparing to bond the carry-through spar onto the rear of the Lynx. Since the carry-through spar supports the load of the wings and transfers that load through the fuselage, the carry-through spar is one of the Lynx's most critical elements. Once the carry-through spar was fitted, the path would open for attaching the strakes, so this represented a major milestone. After many days spent aligning the spar and the fuselage, the engineers and composite technicians successfully bonded the element in place. Now all they had to do was load it to simulate the forces the structure would be expected to withstand during re-entry.

Cockpit

The Lynx cockpit sits inside the outer shell of the vehicle and is attached to the strakes, the fuselage, and the nose. To protect its occupants against heat and debris, the Lynx sports an outer windshield and, to ensure the pressure is maintained inside the cockpit, there is an inner windscreen. Cockpit pressure is maintained at around 2,500 meters, which is a little higher than the cabin pressure of your average commercial airliner. Of course, in an airliner, you're not wearing a spacesuit, which makes for a cozy seating arrangement. Still, if you've flown commercial, you'll be used to being snug in your seat! The flight panel (Figure 4.17) looks very similar to the flat panel instrumentation system found in most corporate jets: pump outlet pressure, checklists, pitch trim, cabin pressure controller – it's all there on the instrument panel.

4.17 Lynx flight panel. Credit: XCOR

As 2014 rolled into 2015, the Lynx was taking shape, since it was now recognizable as a winged vehicle. During the first half of 2015, engineers worked to install electrical wiring and integrated subsystems, and prepared to fit the landing gear bays.

ROCKET POWER

Powering the Lynx will be the full piston-pump-powered XR-5K 18 rocket engine, fueled by liquid oxygen and kerosene, but, before we describe this propulsion system, you may be wondering how rocket engines get their alphanumeric designations and names. First of all, there is no convention that dictates how a company names its rocket engines. Take SpaceX for instance. Every item of hardware in the SpaceX inventory has a name as opposed to a sequence of numbers and letters. There's the Merlin engine, the Falcon launch vehicle, and the Raptor engine. XCOR meanwhile has opted for the use of abbreviations and numbers signifying thrust class and fuel. So, in the case of the XR-5K 18, the "XR" stands for "XCOR Rocket," the "5" signifies the thrust class, the "K" stands for kerosene, and the number 18 represents the fact that this engine is the 18th one that XCOR has designed since the company's inception. Pretty straightforward really. And so back to the XR-5K 18, which you can see in Figure 4.18.

Each XR-5K 18 engine produces 12.9 kN (2,900 pounds of force) of vacuum thrust and the Lynx has four of them. It's an engine that has been thoroughly tested over the years, with the first hot fire test being performed in December 2008. Much of the design shares a heritage with the 4K 14, which logged an extensive flight history(40 flights) as the power source for the X-Racer. Thanks to XCOR's spark torch system, the XR-5K 18 can stop and then be restarted, and thanks to its regeneratively cooled system, the engine can run

4.18 Lynx engine test. Credit: XCOR

indefinitely without the need for time-consuming maintenance. And XCOR has been running the engine regularly. The XR-5K 18 first hit the news in March 2013 when it was mated to the Mark I Lynx fuselage and run for 67 seconds. The test marked the first time a full piston pump-powered rocket engine had been fired. Why piston pumps? Ease of use is the answer. Turbo-pumps are notoriously twitchy and tend to be very expensive, but piston pumps are the Corolla of engines: reliable, affordable, and about as easy to maintain as your average Corolla engine.

The Reaction Control System (RCS)

While the XR-5K 18 engine will provide the power needed for the Lynx's to attain their mission altitudes, the vehicle needs another means of maneuvering in the upper reaches of the atmosphere and in space. As the climb to suborbital altitude continues, the wings and lifting surfaces of the Lynx will rapidly lose effectiveness, which will mean it will be time to switch on the RCS (see Sidebar). At this point, the pilot has to transition from thinking "aircraft" to thinking "spacecraft." It's a challenge that was faced by the X-15 pilots such as Joe Walker, who flew suborbital flights back in the 1960s.

As the Lynx prepares to return to Midland International, the pilot will use the RCS thrusters to stabilize the attitude to prevent the Lynx from entering the atmosphere in a spin or roll condition, which would be bad news for everyone. With the Lynx stable and sinking like a stone, the airflow will build up rapidly and, at around 60,000 meters altitude, the aerodynamic surfaces will once again become responsive and the pilot will switch back to "aircraft" mode, using the aerodynamic controls to control the Lynx's approach to landing.

The Lynx's RCS – the 3N22 – is a bi-propellant thruster, which means it uses a fuel and an oxidizer to do its job. The system has been fired hundreds of times and the system's spark torch igniter has been fired up thousands of times, so you don't have to worry about the RCS not starting! The Lynx is fitted with 12 3N22s in clusters of two (on the nose, on the engine cowling, on two sides of the nose, and on the wing strakes), each cluster being fed by separate feed systems to ensure system redundancy. Building RCS thrusters is something that XCOR has plenty of experience with, having built the XR-3E 17 for military applications, and the XR-3M9, which was tested under contract with the Air Force, and the XR-3B4, which was developed for the National Reconnaissance Office.

5

Spaceports

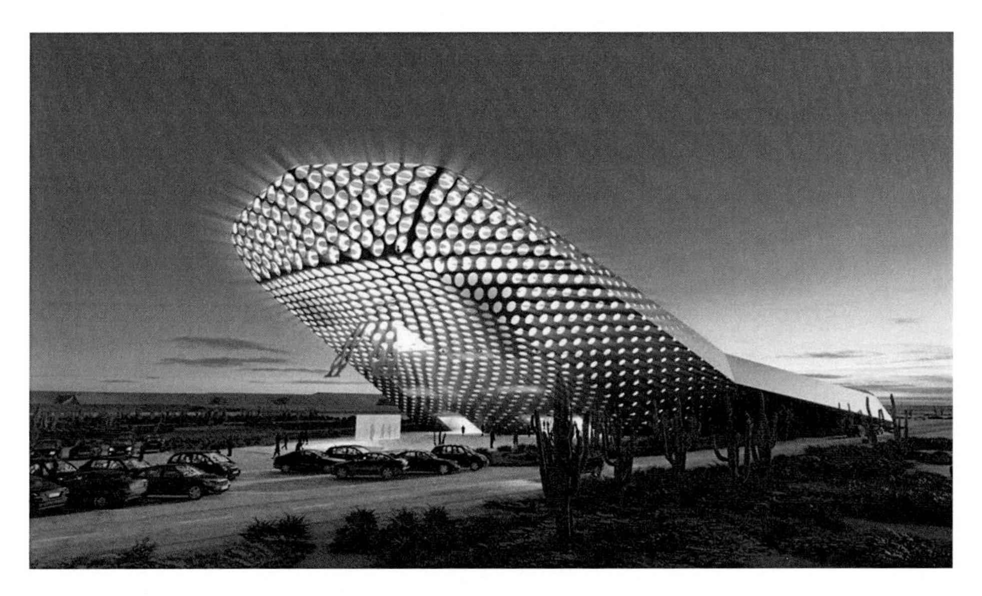

Credit: Caribbean Spaceport/Spaceport Partners

MIDLAND DEVELOPMENT CORPORATION

109 North Main (2nd Floor), Midland, Texas 79701
T: 432.686.3579 or 432.687.8214. Fax 855.824.6435
www.midlandtxedc.com
Executive Director: Pam Welch: 432.686.3552. E-mail: *pwelch@midlandtxedc.com*

E. Seedhouse, *XCOR, Developing the Next Generation Spaceplane*,
Springer Praxis Books, DOI 10.1007/978-3-319-26112-6_5

If you're planning on flying XCOR in the next few years, then you'll either be driving or flying to Midland, site of Midland International Airport, also known as Midland International Air and Space Port. The announcement that XCOR would be relocating their commercial space development center to Midland was made in July 2012, but it took some time for that plan to come to fruition because the airport had to wait for the Federal Aviation Administration (FAA) to grant them a spaceport license. Like so many applications to federal government entities, the spaceport paperwork was complicated because the applicant – Midland International Airport in this case – had to comply with the many, *many* requirements described in an 88-page document entitled *14 CFR Parts 401, 417, and 420 Licensing and Safety Requirements for Operation of a Launch Site; Rule*. Among the requirements was the submission of an environmental assessment plan identifying flight corridors for the spacecraft that would be operating from the runways, developing a plan for handling propellants, another plan to deal with accidents, and myriad airspace plans certification checklists. To give you an idea of just how involved the process was, consider the case of the lesser prairie chicken (Figure 5.1).

Now you may be wondering what on Earth a chicken has to do with a spaceport license but, if you've read up to this chapter, you may recall the story of Virgin Galactic and their battle with the desert tortoise. Well, Midland Airport had a similar battle with a chicken. Visually there is nothing special about your typical lesser prairie chicken, but the feature that makes this bird distinctive is that it is listed as threatened under the Endangered Species Act. It seems these chickens (*Tympanuchus pallidicinctus*, to give them their

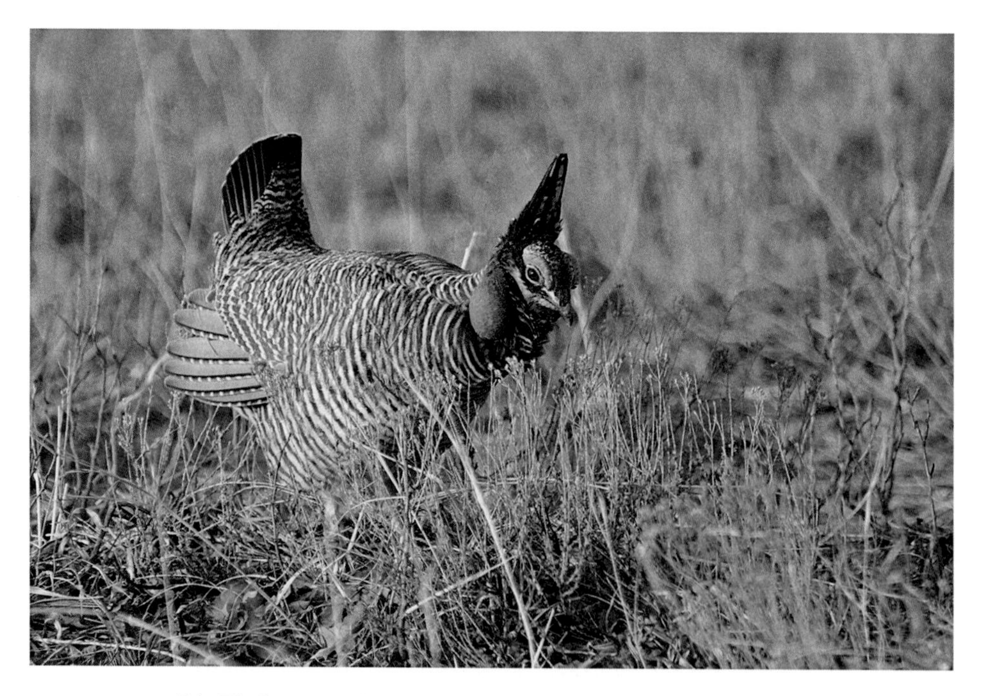

5.1 The lesser prairie chicken, New Mexico. Credit: Steven Walling

proper name) are at risk because their habitat is vulnerable, which meant Midland Airport had to submit an addendum to their spaceport application explaining why the launching and landing of spacecraft wouldn't harm the chickens. Unfortunately, the US Fish and Wildlife Service wasn't appeased, worrying that the sonic booms might disrupt the chickens' mating habits. In response, Midland Airport offered to deploy biologists to study the effect of the first five launches on the chickens – a move that wasn't necessary after the FAA approved the environmental assessment and, in September 2014, issued Midland with its spaceport license.

> "I'm very excited and relieved at the same time knowing we passed a major milestone and made history that is something to be very proud of. The proximity of the airport to the spaceport allows us to take advantage of existing infrastructure, which in turn lowers cost to the operators and offers us a competitive advantage over operations at remote locations."
>
> *Marv Esterly, city of Midland's director of airports, on receiving the license during an FAA Commercial Space Transportation Advisory Committee meeting*

The awarding of a spaceport license to Midland International Air and Space Port (Figure 5.2) meant there were three spaceports all quite close to one another, the other two being Spaceport America in Sierra County, New Mexico, and Blue Origin in West Texas. One element the spaceports had in common was that each had only one anchor tenant and

5.2 Midland International Air and Space Port. Credit: Blueag9

the reliance on just one operator was brought into focus at the end of October following the SpaceShipTwo tragedy. All of sudden, the fanfare of Virgin Galactic revenue flights starting in 2015 came to a grinding halt and pressure was ratcheted up a notch for Spaceport America to find more operators to generate revenue to keep the doors open. The prognosis was grim even before the SpaceShipTwo accident because, the day before the fatal incident, Christine Anderson, Spaceport America's Executive Director, had admitted that the spaceport would be US$1.5 million in the red if SpaceShipTwo didn't start flying in 2015. At the tail end of 2015 it looks unlikely that revenue flights will start before 2018 or 2019, which means that New Mexicans are staring at a US$212 white elephant. It was a lesson to all the other spaceports around the world: put all your eggs in one basket and suffer the consequences. And those consequences could be dire for Spaceport America (see sidebar) as the commercial suborbital spaceport industry gathers steam. As Spaceport America struggles to entice other operators such as World View Enterprises, there may be steps to sell the facility as patience about the promises made by Virgin Galactic wear thin. That was the story in February 2015 when the Senate Corporations and Transportation Committee moved bill SB 267 to the Senate Finance Committee – a move that could kick-start the sale of the unused spaceport. But who will buy it?

Spaceport America

Opened to much fanfare in October 2011, little has gone right for those involved in the New Mexico spaceport project. Located in an area known as *Jornada del Muerto* (route of the dead man), Spaceport is inconveniently located some 140 kilometers north of El Paso, the nearest airport (no commercial flights serve Spaceport America).

www.forbes.com, Alex Knapp, 20 February 2015

MIDLAND INTERNATIONAL AIRPORT

Midland International Airport (FAA and IATA Code: MAF) is a primary small-hub airport certified by the FAA under Federal Aviation regulation (FAR) 139. And, since September 2014, it has also been certified as a spaceport under FAR 420. If you happen to be planning on flying on the Lynx, you'll be pleased to know that MAF (see sidebar) is served by American Eagle, Southwest Airlines, and United Express, with more than 20 flights daily, serving Dallas, Houston, Las Vegas, and Denver. No driving through the wilderness to visit this airport. And, if you're bringing family and friends, there are plenty of hotels, car-rental facilities, and restaurants. For XCOR, the location (Figure 5.3) is ideal for launching the Lynx. Launch and landing corridors have been established and the climate offers 315 VFR (visual flight rules) days every year so the chances of a flight being delayed due to weather will be remote.

5.3 Midland International Airport (MIA). Credit: Midland Development Corporation

Midland International Airport (MAF) by the Numbers

Location: 31°56′33″N 102°12′07″W
Elevation: 875 meters
Website: *www.FlyMAF.com*
Runways:

4/22: 1,404 meters
10/28: 2,530 meters
16L/34R: 1,323 meters
16R/34L: 2,896 meters

First commercial spaceport co-located with a commercial airport
Busiest domestic routes (June 2013–May 2014):

Rank	*City*	*Passengers*	*Carriers*
1	Dallas–Fort Worth, Texas	189,000	American
2	Dallas–Love Field, Texas	92,000	Southwest
3	Houston–Intercontinental, Texas	69,000	United
4	Houston–Hobby, Texas	52,000	Southwest
5	Las Vegas, Nevada	27,000	Southwest
6	Denver, Colorado	1,000	United

In addition to XCOR, Midland International Air and Space Port's Spaceport Business Park is host to Orbital Outfitters, a company that provides a range of services to the commercial spaceflight industry, including the manufacture and design of space and pressure suits, and performing high-altitude testing in their chamber complex. In anticipation of a steady stream of passengers, MAF has completed a master plan for their Spaceport Business Park that includes the construction of new roads and ground leases for aerospace development.

INTERNATIONAL SPACEPORTS

Flying from MAF will be all well and good if you happen to live in North America, but what happens if you live in Europe or Australasia and you want the Lynx experience? Well, there are other spaceports in the works and we'll discuss some of the likeliest candidates here. Of course, for the Lynx to fly from any of these potential spaceports, there is the not-so-insignificant issue of the export license that will be required to allow the weapon – sorry, spacecraft – out of the US. While XCOR assure their customers that this is a formality, don't forget we're talking about the government here, so it is still an administrative bridge that must be crossed. Still, let's hope for the best.

Australia

So Midland International Air and Space Port is probably where the suborbital action will be for at least the first few years, but where else might the Lynx take off from? Well, one possibility is Australia. The man leading the charge is John Moody, a musician and producer, who has identified three sites in Queensland: Townsville, Rockhampton, and Toowoomba. Of the three, one likely candidate would be Townsville thanks to its 3,000-meter-plus runway, although Rockhampton is also favored thanks the town's proximity to the state capital, Brisbane. By 2015, Moody had been working on the project for almost three years and had lobbied the State Government and various corporations with vested interests in commercial space development. He also met with XCOR President Andrew Nelson and persuaded ex-NASA and Defense Advanced Research Projects Agency (DARPA) contractor Ethan Chew to sign on as Chief Technical Advisor.

Scotland

Not a big fan of endless sun and blue skies? Then head to Scotland, where the British Government is also touting plans to build a spaceport. Locations under consideration include RAF Lossiemouth, Glasgow Prestwick, Stornoway Airport, and Kinloss Barracks (there are also a couple of locations being considered in the south of England). The decision to go ahead and build a spaceport in the UK goes back to a query in 2012 made by the UK Department of Transport and the UK Space Agency to the Civil Aviation Authority (CAA) asking how Britain could make the most of the impending suborbital space industry. The CAA responded by issuing a report that stated suborbital operations could be a reality as early as 2018 and, all of sudden, airports around the country started promoting their runways as potential spaceports. The report also admitted that a Scottish location might not be the best place given the strong winds, cloud cover, and lots *and lots* of rain, all of which would translate to fewer flying hours. But the CAA's report didn't seem to affect Scottish enthusiasm for locating Britain's spaceport in the north because, by mid-2015, it was Prestwick that was the frontrunner in the spaceport race. Although the weather can't compete with locations further south, Prestwick is well connected by road and rail, and it also happens to have one of the longest runways in Scotland. Locating a spaceport there would also help the ailing loss-making facility which, in 2014, was losing more than a million dollars a month.

Curaçao

In case Australia is too far and Scotland too cold, how about a nice trip to the Caribbean? Curaçao perhaps? The concept of a Caribbean spaceport has been around for a while now, with the idea first conceived in 2005 by a group of investors that combined to form Spaceport Partners. Since 2005, Caribbean Spaceport has worked with a variety of government, academic, and business entities to determine the technological and economical viability of operating a spaceport in the Netherlands Antilles. Among the entities are the Netherlands Ministry of Economic Affairs, Delft University's Faculty of Space and Aerospace Engineering, Remco System Construction, and DDOCK Design Development. The US$1.8 billion concept (Figure 5.4) calls for a 200-hectare area of land to be turned into a spaceport from which the Lynx will operate under that wet lease agreement mentioned earlier.

5.4 Caribbean Spaceport. Credit: Caribbean Spaceport/Spaceport Partners

Why Curaçao? Why not? The locals are very friendly and accommodating, and the location has long been a popular tourist destination for the reasons described on the XCOR website:

> "The Caribbean offers magical blue waters and white sandy beaches, against a bright green backdrop. Seen from space, it is absolutely spectacular. For those traveling with you to Curaçao, but not into space, explore the town of Willemstad, which is on the UNESCO world heritage list. The deep-sea submarine, fine beaches and great diving are a Caribbean treat to make their stay unforgettable."

Travel agents seem to think it's a great idea, with some already offering this once-in-a-lifetime experience. Merit Travel, for instance, posts this enticing ad on their website:

> "The Astronaut Program taking off from Curaçao, the Space Expedition Corporation home base in the Caribbean, is for all enthusiasts eager to travel to space and experience this once-in-a-lifetime adventure. By booking your flight into space you become one of a select group of Canadians to have shared this space voyage. Your space flight will take you up 103 kilometers above earth where you will enter low orbit, officially crossing the frontier of space. The flight includes airfare to the spaceport of your choice, a 3 day stay in a five star hotel, individual mission training and a documented video and photo log of you crossing the frontier of space. Ultimate bragging rights come standard. After reaching orbit you can officially give yourself the title 'ASTRONAUT.'"
>
> *Source: www.merittravel.com. T: 1-866-341-1777*

6

Missions and Payload Integration

Credit: XCOR

"Lynx is revolutionizing space-based research and making space easily accessible to the AGU community via its '*Your Mission. Our Ship.*' program for customized scientific and education payload flights. The spacecraft will offer high frequency flights entirely dedicated to the researcher's mission. This low-cost platform provides the ability to design and conduct repeatable, high resolution experiments. Scientists will

E. Seedhouse, *XCOR, Developing the Next Generation Spaceplane*,
Springer Praxis Books, DOI 10.1007/978-3-319-26112-6_6

be able to gather in-situ measurements at multiple points in the atmosphere, conduct solar and space physics research, or direct a human-tended telescope at planetary objects for above-atmosphere astronomy."

Khaki Rodway, XCOR's Director of Payload Sales and Operations, speaking at the American Geophysical Union's annual meeting, December 2014

SCIENCE AND PAYLOAD MISSIONS

If you have been following the suborbital space game over the years, you will no doubt be familiar with the glitzy Virgin Galactic presentations featuring space tourists floating around SpaceShipTwo's cabin clad in their cool flight suits. That's because SpaceShipTwo was designed primarily with thrill-seeking tourists in mind, although, if you happen to be a scientist, Virgin Galactic can configure the cabin for payloads to be flown and experiments conducted. When it comes to XCOR, the emphasis has mostly been on science, partly because the sole passenger in the Lynx will remain strapped in for the duration of the flight. At the AGU meeting in 2014 (see the above quote), prospective suborbital scientists were given the opportunity to examine payload experiments and discover for themselves just how versatile a platform the Lynx will be for flying science.

The science and payload potential of the Lynx first attracted widespread attention at the 2011 Next-Generation Suborbital Researchers Conference (NSRC) in Orlando, Florida. It was in Orlando that XCOR announced its team of payload integration specialists that included the Southwest Research Institute (SwRI), which announced it had bought six tickets on the Lynx, and NanoRacks, which had already flown research platforms on the International Space Station (ISS). Then there were companies such as Andrew Space and Spaceflight Services, who planned to use the Lynx as a platform to test experiments prior to them being flown on SpaceX's Falcon 9 to the ISS. All the details of how experiments would be flown and how payloads would be integrated would be explained in XCOR's *Payload User Guide* (PUG).

One of those payloads will be the Atsa Suborbital Observatory (Figure 6.1), a telescope designed to provide a low-cost space-based observation facility thanks to a deal inked in July 2011 between XCOR and the Planetary Space Institute (PSI). Developed by Atsa Project Scientist Faith Vilas and PSI affiliate scientist Luke Sollitt, the suborbital observatory has been designed to observe objects near the Sun that can't be studied using traditional orbital telescopes. One advantage of the Atsa is that it will be able to fly on a customized flight trajectory and, since it will be human-tended, the observatory can be pointed at specific targets of interest. And, since the Lynx will be capable of flying several times a day, the Atsa can be used to investigate targets of opportunity.

In addition to the obvious scientific benefits, the Atsa program has an educational component, since one of the goals of the program is to have students operating the observatory. To that end, operation simulation and training are being conducted at PSI's headquarters in Tuscon, Arizona, to help students learn how to work and fly in space. And they will need that training because Atsa operators will be encumbered by a pressure suit and will have only three or four minutes to conduct their observations. To get an idea of the challenges

6.1 Lynx with the Atsa Suborbital Observatory. Credit: XCOR

of manipulating and operating equipment in the cozy confines of the Lynx, students have completed "fit and function" tests using the Lynx mock-up. Future work includes testing the camera under G using NASTAR's centrifuge and on board parabolic flights.

FUNDING OPPORTUNITIES: A BRIEF HISTORY

Shortly after the Atsa deal, XCOR received another shot in the arm when NASA announced the company had been selected to provide payload integration services for research missions in a program that would provide as much as US$10 million in contracts. In awarding XCOR the contract, NASA had recognized the Lynx for the versatile platform that XCOR was designing the vehicle to be. Thanks to the contract, the agency was also incentivizing and encouraging low-cost access to space in what has long been an underserved market due to the long lead times on other suborbital platforms (such as sounding rockets, for example). The contract award was under NASA's Flight Opportunities Program (FOP), which is incorporated into the Space Technology Program, which is in turn managed by the Space Technology Mission Directorate (STMD). The goal of the Space Technology Program is to serve as NASA's technology and demonstration incubator, working with education institutions, industry, and other government agencies to develop space capabilities by testing in space-specific environments such as zero-G and on board suborbital platforms. The FOP, which is just one of nine STMD programs, was established in 2010, but its origins go back to September 2008 when NASA first used commercial microgravity

6.2 2010 FAST Flight Week with Embry-Riddle Aeronautical University students. Credit: NASA

flight services. In September that year, the agency had sponsored five companies under its Facilitated Access to Space Environment for Technology Development and Training (FAST) – to fly experiments on board an aircraft operated by Zero Gravity Corporation. Then, three months later, the Universities Space Research Association (USRA) sponsored a workshop attended by representatives of NASA's Ames Research Center (ARC) and the Commercial Spaceflight Federation (CSF) to discuss the possibilities offered by the impending commercial suborbital vehicles. This workshop was followed by another held in conjunction with the Aerospace Medical Association's Annual Meeting, after which NASA selected technology demonstration projects to be flown on reduced-gravity flights through its FAST program (Figure 6.2).

The following year, NASA established the Commercial Reusable Suborbital Research Program (CRuSR) under its Innovative Partnerships Program (IPP) with the intent to procure reusable suborbital spaceflight services and to solicit research studies to utilize those services. Then, in the first half of 2010, NASA selected another batch of technology demonstration projects for reduced-gravity flights. This announcement was followed by the agency hosting the Space Technology Industry Forum to discuss new space technology investments – an event that focused on the 2011 budget for new Space Technology Programs. 2010 ended with NASA announcing that it was seeking proposals from scientists interested in testing new technologies during suborbital flights – a step that set up a meeting the following year for potential providers of suborbital vehicles. Later in 2011, NASA, as part of its FOP, selected 16 payloads for flights on board Zero-G's parabolic aircraft and two suborbital vehicles. Shortly after the selection, the agency made another announcement for proposals for suborbital services and then, in August, NASA announced the aforementioned two-year contracts worth up to US$10 million. The following year saw the suborbital commercial spaceflight industry gather more momentum when NASA's FOP selected 24 payloads for flights on an assortment of reduced-gravity platforms, including five slated to fly on suborbital vehicles. 2013 was much the same as 2012, with the agency, in what was now its sixth cycle of selections, selecting more than 30 space technology payloads for suborbital flights. By the close of 2013, more than 100 technologies with test flights had been procured through NASA's STMD's FOP. While 2013 had been the STMD's busiest year, 2014 promised to be busier, with five parabolic flight campaigns, two suborbital launches courtesy of UP Aerospace, and an assortment of other reduced-gravity flights. With the April announcement of another 13 space technology payloads for flights on board commercial suborbital vehicles, NASA had sponsored 130 flight opportunities – a number that was boosted the following month when it was announced that the agency had selected a dozen technology experiments on the first commercial research flight of Virgin Galactic's SpaceShipTwo.

In a business where it costs US$10,000 to loft just one kilogram into orbit, the FOP is a godsend to those who want to test in extended periods of microgravity. Without FOP, many scientists and researchers would be left twiddling their collective thumbs, grumbling about the high costs of developing the hardware and operational capability required to fly their payload in low Earth orbit (LEO). At the same time, due to the prohibitive costs of flying to LEO, many payloads just can't be flown, which results in many technologies simply sitting on the shelf while the workforce that could have gained experience working on those technologies sit under-utilized and untrained. So the FOP (see sidebar) bridges a crucial gap between cutting-edge technology and mission-specific operational environment thanks to the new breed of suborbital vehicles.

The Flight Opportunities Program (FOP) Commercial Flight Providers[1]

Armadillo Aerospace, Heath, TX
Near Space Corp., Tillamook, OR
Masten Space Systems, Mojave, CA
UP Aerospace Inc., Highlands Ranch, CO
Virgin Galactic, Mojave, CA
Whittinghill Aerospace LLC, Camarillo, CA
XCOR Aerospace, Mojave, CA

---

1 These flight providers are at various stages of development. Some are still in the design and development phase, while others, such as XCOR, are approaching flight testing.

PROJECT POSSUM

Perhaps the best-known payload and research idea to be funded through NASA's FOP is Jason Reimuller's Project PoSSUM (see sidebar) (Figure 6.3), an acronym that stands for "Polar Suborbital Science in the Upper Mesosphere" (see Chapter 9). Project PoSSUM (Appendix IV), which was known in an earlier life as the Noctilucent Cloud Imagery and Tomography experiment, was selected by the FOP in March 2012 as experiment 46-S. One particularly unique feature of the Project PoSSUM venture (*www.projectpossum.org*) is that it is the first manned suborbital research program, and another distinguishing aspect is that the program is twinned with the world's first science-based suborbital astronaut training program, which will be discussed later.

Project PoSSUM

At the core of Project PoSSUM is advancing our understanding of noctilucent clouds (Figure 6.4), which are also known as Polar Mesospheric Clouds (PMCs), which is how Jason came up with the memorable PoSSUM acronym. PMCs are not your common cloud, typically forming at altitudes of between 80 and 85 kilometers, which happens to be just a few kilometers below the coldest division of the atmosphere, the *mesopause*. Comprising very small ice crystals just one-tenth of a micron in diameter, PMCs can be seen thanks to sunlight scattering by the crystals. For PMCs to form requires a combination of extremely low temperatures, water vapor, and nuclei. Why all the fuss about PMCs? Well, these clouds happen to be of interest to those working in the aeronomy and climate science arenas because PMCs are sensitive indicators of what is happening in the upper atmosphere. Also, the brightness of these clouds has increased over the years and they have extended to lower latitudes, but scientists aren't sure why.

6.3 Project PoSSUM mission patch. Credit: Project PoSSUM

6.4 Noctilucent clouds. Credit: NASA

PAYLOAD INTEGRATION

Many scientists and researchers will be unfamiliar with the payload integration process, so what follows is a primer that provides a generic overview of the roles and responsibilities of the Principal Investigators (PIs) and Payload Integration Manager (PIM). This primer also describes the inputs required by the PI and the services available to the PI during the payload development process. We'll begin with the beginning of the payload process (Figure 6.5) and work our way step by step to the manifest payload and finally to the integration and launch of that payload. The core focus of this section is to provide the PI with the tools needed to develop their product so it's ready for flight on board the Lynx. To that end, we'll stay clear of as much of the technical jargon as possible, with the aim of making this readable to those new to the world of payload integration.

The Payload Integration Process

In this section, we'll outline the steps required in the payload integration process (Figure 6.5). The first step in the process is the Pre-flight Phase, which defines the payload requirements, the payload design, safety reviews, and the flight assignment, which is when the payload makes its way onto the Lynx manifest. This phase also drives the crew training, which in most cases will include the PI and his or her designated backup. The second phase is the Flight Phase, which includes the process of payload integration, the flight of

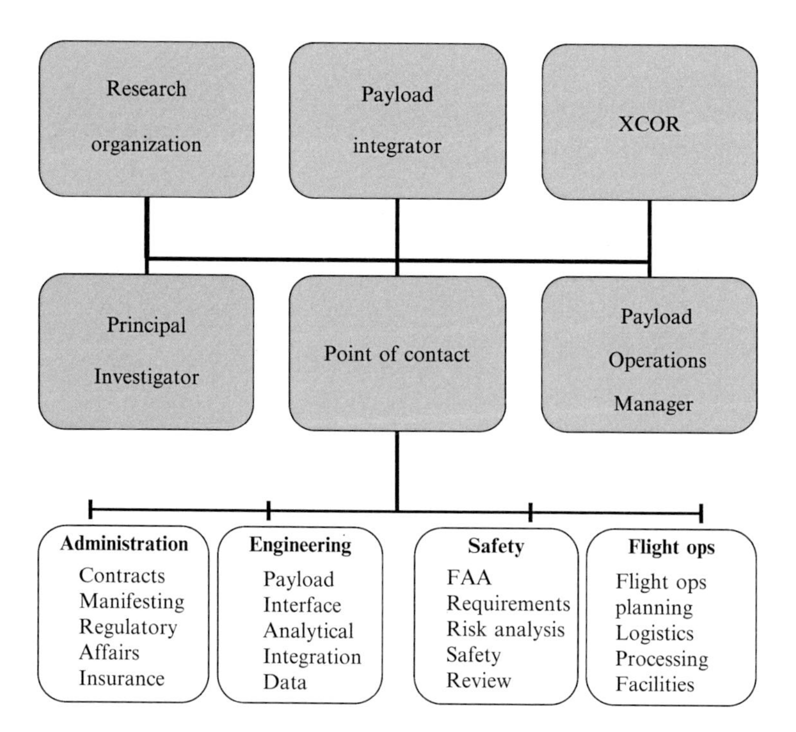

6.5 The payload integration path. Credit: Author's work

that payload, and the return of that payload. The Post-flight Phase includes tasks such as de-integration, the return of the payload from the landing site (which in most cases will be Midland Airport), lessons learned, a crew debrief, and the mission report.

The payload integration process for these missions will be referred to using the following terms: "Payload minus date" (P-XX), "Flight minus date" (F-XX), and "Return plus date" (R+XX). The "Payload minus date" defines how many days a task must occur before the payload must make its way to the PIM. For example, P-40 indicates that the PI has 40 days to get his or her payload to the PIM. A "Flight minus date" indicates how many days a task must be completed before launch and a "Return plus date" indicates when tasks begin following landing. It should be noted that some of the payload integration tasks proceed sequentially but others may proceed in parallel. It just depends on the complexity of the payload. A typical payload integration flow will follow a fairly specific timeline, beginning with the assignment of the PIM, after which the following steps will occur:

- Payload performance
- Payload manifest
- Hardware interface development
- Software development
- Human factors
- Payload safety review
- Operations integration

- Testing
- Hardware delivery
- Bench review
- Certification of flight readiness
- Payload launch-site support
- Launch
- Suborbital operations
- Landing and return of payload
- Post-flight tasks.

Payload Development

So you have a payload and/or science experiment you would like to fly on the Lynx. What now? Well, there's a fair amount of work that's needed before your project can be launched to the edge of space and this section explains what you will need to do. But first read XCOR's PUG. Done? Great. Now, if you're flying an experiment, you will need to complete an experiment summary together with details of the payload resource requirements because an experiment is also classed as a payload. You will also need to make a list of the activities that must be performed from launch to landing, details of baseline data collection for the investigation, and any stowage needs. If your investigation is a life sciences experiment using a human test subject, you will need to identify the tests and evaluations to be performed pre flight, in flight, and post flight, together with reference points (if any). To give you an idea of what is involved, I've included a test template in Appendix V for a life sciences experiment together with an engineering template which is more payload-oriented.

If your experiment involves testing a human subject, you will probably need to conduct baseline testing which will involve a series of pre-flight tests to establish reference points for comparison with in-flight and post-flight data. We'll refer to this as Baseline Data Collection, or BDC. At some point in the lead-up to the flight, the BDC will be reviewed by the mission team members and XCOR representatives so everyone has a good understanding of the investigation's requirements and constraints (power requirements, data and downlink needs, photographic requirements, and stowage needs, for example). For simple protocols such as measuring blood pressure, it will simply be a matter of referring to the relevant protocols, such as those in Appendix VI, but, for more complex investigations, conflicts will no doubt be identified and proposals to solve them will be developed. Once the conflicts have been resolved, you need to start thinking about the Informed Consent Briefing (ICB). If you as the PI happen to be the test subject, this matter won't be a problem! The ICB will then be submitted to XCOR and you can move on to the mission integration process.

Mission Integration

Most of the mission integration tasks are dealt with by the PIM, who has myriad roles and responsibilities in this regard. Once you have been assigned a PIM, he or she will be your primary point of contact for all matters pertaining to the payload you plan to fly. The PIM will be responsible for integrating your payload into the Lynx and also for developing the payload Interface Control Document (ICD), the payload schedule, and any payload data products. They will also manage the data deliveries in support of your mission and ensure that your payload requirements are correctly defined, documented, and compatible with the Lynx. If changes to your payload need to be made then the PIM will ensure these changes are captured in the Change Evaluation Form (CEF). Sometime in the mission integration process, the PIM will arrange a Hardware Feasibility Assessment (HFA) to ensure the viability of the payload and to assess payload readiness. The next stage will be a flight readiness review (FRR) to ensure your payload is safe to be flown and, assuming the PIM can make checks in all the necessary boxes, a Certificate of Flight Readiness (CFR) will be issued. With this piece of paper in hand, you are now ready to prepare for payload delivery to the launch site, which will most likely be Midland International Air and Space Port. While the PIM can advise you on this task, the business of getting your payload to the launch site is your responsibility. Ideally, the payload should arrive three or four days before the launch date to allow for processing of forms and documentation and also to allow sufficient time for a crew equipment interface test if required. This test, which will most likely be conducted in the Lynx mock-up, provides the crewmember with the opportunity to verify hardware interfaces, functional testing of the payload, and hands-on internal and external verification of the payload in-flight configuration. It is also an opportunity for XCOR representatives to inspect the payload one last time to make sure it meets all the requirements for flight on board the vehicle.

Engineering Integration

Another of the PIM's myriad responsibilities is helping you ensure that your payload conforms to XCOR's interface requirements. Normally this is a very straightforward process but, if a payload deviates from the specified requirements and/or interfaces, then an exception submission must be issued. This paperwork, which will usually be accompanied by supporting documentation, includes details of the specific deviations from the standard requirements and also a rationale for its acceptance. After reviewing the documentation, a waiver may be issued allowing the payload to be flown or, as is more likely, the payload will need to be modified so that it conforms to payload regulations and flight rules.

Another task under the engineering integration umbrella is that of human factor integration. As with most matters pertaining to payload integration, human factor integration is the PIM's job. To ensure the payload is safe to fly with crew, the PIM will ensure it meets the requirements for touch, temperature, noise, vibration, and other factors that affect human performance. If your payload fails to comply with the human factor integration requirements, it will be necessary to modify the payload to ensure compliance.

Payload Software

There are all sorts of software that can be used to ensure payloads operate as advertised, and the type of software you choose will depend on factors such as data handling, telemetry, and monitoring operations. Bear in mind that you will be transporting your payload to a remote location – the spaceport – and you will need to perform system and compatibility testing at that location, which means all applications should be capable of being run from a laptop. You may also need to interact with your payload during the flight or have the astronaut interact with the payload, so these factors should be considered when designing the display. Given the cost of the mission (US$150,000 divided by four minutes equals US$37,500 per minute or around US$625 per second), you should spend as much time as necessary to ensure flawless software operation – a task that should be performed at every stage of the payload development cycle. If you do this, then the chances are good that your payload will be error-free and your mission will be successful.

Payload Operations

Once the payload is developed, the PI/astronaut can get on with the business of training, which requires a training strategy to be formulated by a crew-training coordinator, which in some cases may be the PIM, depending on their background. Whoever plans the training must take into consideration factors such as displays, procedures, and payload familiarization. At this point, it is important to note that we're talking about payload-specific training, since general astronaut training will be provided by XCOR on site at the Midland International Air and Space Port. To develop the payload-specific training, the PI and the PIM will sit down and sift through the mission's procedure milestones, most of which will need to be slotted into those very precious four minutes. As part of the payload-specific training, the PIM, in consultation with the PI, will develop a crew payload procedure manual which lists the sequential tasks of the mission along with a timeline of events. Some procedures listed in this manual may simply require flipping a switch, turning a camera, or calibrating an instrument. No matter what the task, each item will be identified in the manual. Given the time constraints of a suborbital mission, the astronaut won't have time to start thumbing through a manual, so the tasks will be presented on a head-mounted display (HMD). The HMD's primary function will be as a procedure viewer capable of rendering step-by-step instructions to the astronaut. It will also be capable of saving critical images and video notes for post-mission analysis. Capable of voice recognition, the HMD will allow the astronaut to focus on the mission in hand without having to worry about looking down at a checklist – as long as the speech-recognition system is designed with mission inputs in mind that is. One of the concerns of using a speech-recognition system in a spacecraft is noise, which can result in misinterpretation, which can in turn result in errors and a very, *very* expensive flight. So it will be the PI's job to "train" the recognition system by creating audio notes specific to the mission tasks. The system should then be tested and retested to ensure the system can screen out and filter misinterpretations caused by the noise of the vehicle and the background noise of the pilot communicating with the ground.

In addition to the flight task list, the manual will also include pre-flight and post-flight procedures together with details of how to respond to off-nominal events. For the broader payload-specific training, the PI will be trained in experiment objectives, mission operations, vehicle familiarization, and crew resource management. Some of this training will take place in the classroom, some in a hands-on environment such as the simulator, and some via courseware. As the PI, think of this as an extension of on-the-job training.

Payload Safety Review

As you approach the launch date, you will be required to prove to XCOR that your payload complies with the technical and safety requirements. Since your PIM will have conducted periodic safety reviews during the development of your payload, the safety review should be a straightforward process, but its best to be prepared all the same. Meeting the requirements of the safety review is all about hazard reduction, so be sure to check your payload is as safe as it can be.

Export Control

If you are a foreign national with a ticket to space, then this section will be particularly important. An export item is anything that has been shipped or transferred from outside of the US; it can be software, a laptop, or a memory device. It doesn't matter: if it originates from outside the US, then it's an export. Once again, it will be the PIM's job to ensure the payload complies with the US Government's Export Control Regulations.

Payload Processing

When you finally reach the stage with all your ducks in a row, you can begin thinking about payload processing. The support requirements for payloads will obviously vary depending on the type of payload and we'll discuss some of these requirements in the payload checklist. Perhaps the most demanding processing requirements are those for biological sciences experiments, since these often not only demand time-sensitive activities such as delivery of samples, but also require consumable supplies, chemicals, biohazard waste management, equipment temperature monitoring, and time-sensitive shipping support. If your payload happens to fit into the biological sciences category, it is worth bearing these factors in mind, just as it is worth checking on the shipping requirements, the acceptance criteria, and the turnover process: do you know what the custodial turnover process is for shipping a biological sciences payload from, say, Vancouver, Canada, to Midland, Texas?

Once your payload has arrived at the launch site, it can be physically integrated into the Lynx. If it is a biological sciences payload, then you will need to check those aforementioned items and also be wary of any flight delay with regard to time-critical or conditioned samples: do you have a plan for late access to the vehicle or if the flight is delayed for an hour? A day? A week? It could happen. Let's hope everything goes by the book and the flight launches on schedule. The Lynx rockets upwards and you follow the progress of your payload on your laptop but, when the vehicle enters the microgravity phase, you notice an anomaly. In fact, there are all sorts of off-nominal behaviors. Nothing that affects the safety of the crew, but the payload is a bust. What do you do? Well, you will need to complete a Payload Anomaly Report (PAR), which fully documents all features of the anomaly based on the available data. This will hopefully allow you to troubleshoot the problem and develop a procedure so that the anomaly isn't repeated. In the early days of flying payloads on commercial suborbital vehicles, it is unlikely that any payload will operate flawlessly so, after each flight, it is worth sitting down and documenting lessons learned. This exercise can be performed in conjunction with the crew debrief if the flight was carrying a passenger. Examples of lessons learned are equipment malfunction, communication failure, hardware/software anomalies, ground operations deficiencies, and misleading crew procedures. For the crew debrief, which should be scheduled within a day of landing ideally, the PI should assemble his or her team who should be accompanied by the pilot. The pilot will first provide an account of how the vehicle performed during flight. He/she will be followed by the PI, who will provide a first-hand account of how the science and hardware performed. This debrief will address issues such as science observation, crew procedures, products, hardware/software performance, and crew training. The length of the debrief will most likely be related to the rigorousness of the pre-flight training, so here are some pointers to help you have a successful flight:

- Plan for contingencies and include plenty of contingency training in your pre-flight preparation
- Most of your contingency planning should be focused on the failures that are most likely to occur or those that are most likely to have the greatest impact on your mission
- Every minute you spend planning before your flight results in a more productive and successful mission
- Flexibility in the plan is paramount – have a planned response for as many foreseen and unexpected situations as possible
- Train, train, and train again for contingencies – this training should be performed in simulators as much as possible.

Two to four weeks following your mission, you should prepare a post-mission report (PMR) that describes what was accomplished. The PMR doesn't include data analysis, but describes what worked and what didn't, any snafus, and solutions to those issues. This report should be sent to your sponsors, any stakeholders, partners, and XCOR.

PAYLOAD INTEGRATION CONSIDERATIONS

So that was a brief overview of the world of payload planning, development, integration, and operations. Now let's move on to some of the more detailed aspects of this process by starting with where you, the PI, may be sitting: the cabin.

Cabin Characteristics

There are number of factors you need to consider (see below) in the cabin, but for many scientists the key factor will be the size. As you can see from Figure 6.6, the Lynx is very snug – a feature compounded by the fact that you will be wearing a spacesuit which makes maneuverability a challenge at the best of times. In addition to considering the elbow room, you will also need to plan for various cabin constraints, including:

6.6 Jonna in Embry-Riddle Aeronautical University's Suborbital Spaceflight Simulator. Credit: Project PoSSUM

- cabin pressure under normal operation;
- cabin pressure under abort operations;
- cabin rate of pressure change under normal operation;
- repressurization/depressurization rate during nominal and off-nominal operations;
- particulate cabin air concentration;
- carbon dioxide and oxygen concentration;
- temperature during nominal operations;
- temperature ranges during all mission phases.

Payload Characteristics

Checked the cabin? Ok, let's move on to your payload. Each Lynx variant will be capable of carrying a primary payload in the right seat. This payload can either be a human wearing a spacesuit or two Space Shuttle middeck lockers (Figure 6.7) stacked on top of one another. As long as either payload doesn't weigh more than 120 kilograms, you're good to go, provided you comply with the restrictions outlined in this section.

When operators are presenting their payload capabilities at conferences, one of the most common questions is whether animals can fly. Most operators reply that such requests will be considered on a case-by-case basis, which suggests that animals might be accommodated, most likely in some form of animal-enclosure module (AEM) (see sidebar).

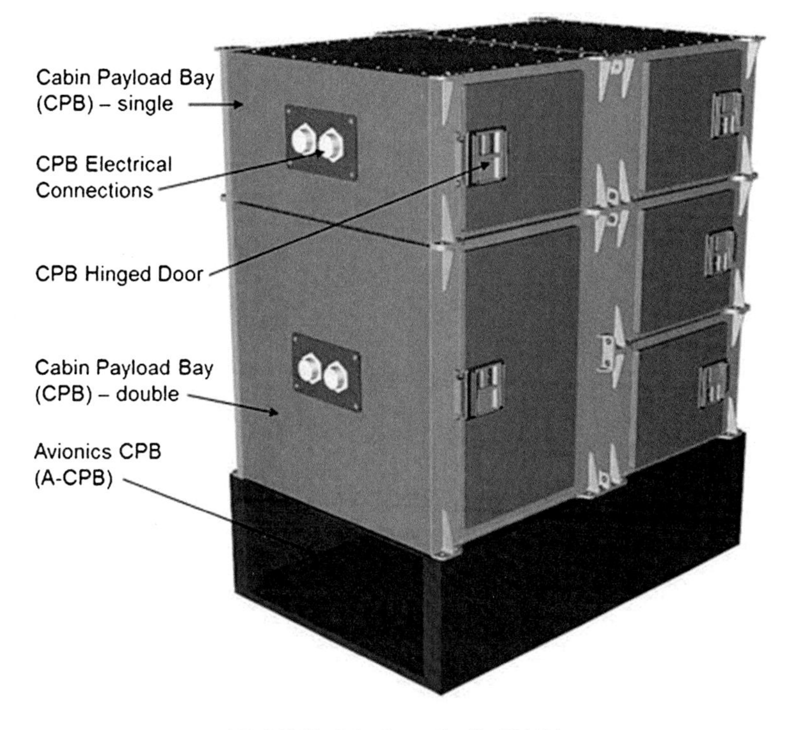

6.7 Middeck lockers. Credit: NASA

Animal-Enclosure Module (AEM)

If you are planning on sending animals into space on board the Lynx, you will need to develop an AEM (Figure 6.8) that slots into the middeck lockers. You will also need to ensure that your AEM is a self-contained habitat capable of providing its occupants with living space, food, water, ventilation, and lighting. And, since your payload may be sitting on the ramp waiting to be loaded, you need to ensure the AEM features an internal waste-management system and a means to prevent any waste from escaping into the confines of the cabin. If you would like to film the activities of your space-bound guinea-pigs, you should ensure that at least one panel of the AEM is Lexan or a similar space-tested material. Another important feature will be temperature. It can get quite hot inside a spacecraft and you want to make sure your test subjects are comfortable, so fitting a few fan blowers won't hurt. You should also ensure that the air quality is maintained within certain limits, which means a high-efficiency particulate air (HEPA) filter should be a feature of your AEM to prevent any microbiological contaminants escaping into the cabin. Water can be provided via drinking valves linked to flexible bladders and food can be delivered in the form of a sterilized laboratory formula. Remember, everything in the AEM needs to function automatically because the pilot won't be able to interact with the payload during flight.

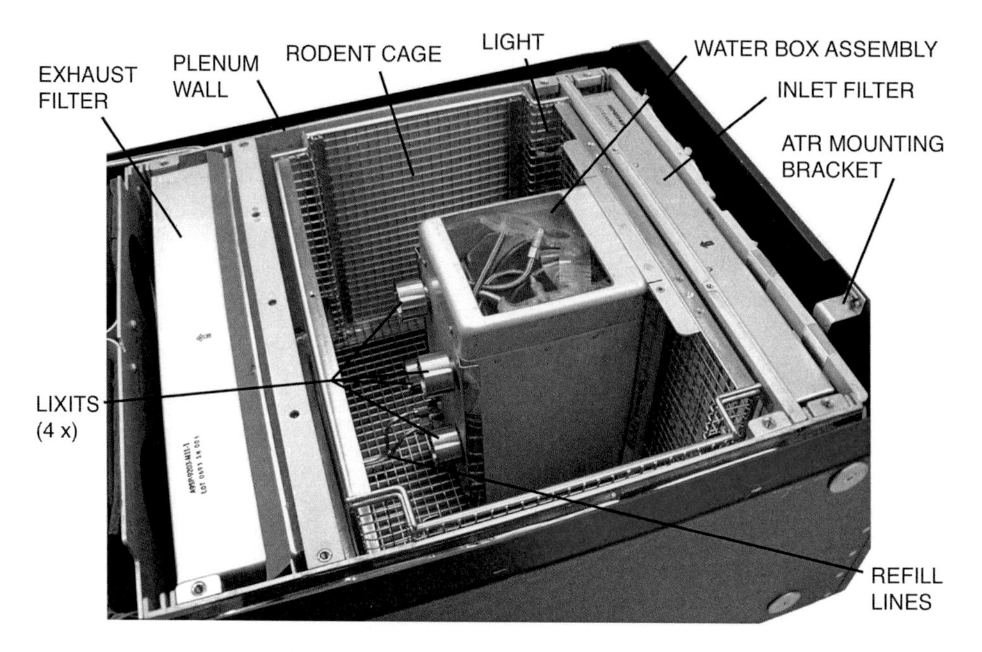

6.8 gAnimal-enclosure module (AEM). Credit: NASA

Structural Interfaces

Regardless of what type of payload you will be flying, you will need to ensure that it can survive not only the launch and re-entry loads, but also the vibration that occurs during a spaceflight. Also, don't forget that the Lynx is maneuvered during the microgravity phase of the flight through the use of the vehicle's reaction control system. While these accelerations are much smaller than those of launch and re-entry, they still represent a design condition for your payload. And, if the flight goes pear-shaped and the Lynx has to execute an emergency landing, your payload must be designed to survive emergency-landing loads. But it is no good designing your payload to survive the minimum acceleration or landing loads: to ensure your payload is accepted at the safety review, it will probably have to have an ultimate factor of safety of at least 2.0 and perhaps as high as 4.0. That means your payload must survive intact – no fracturing or structural failure – at four times the Lynx's acceleration and landing loads. How do you ruggedize your payload to survive these loads? Well, you can use distortion-tolerant foam padding and you can fix net retention devices/interfaces. And, when you've done as much ruggedizing as you can, you can test. And test again. And again.

Environmental Conditions

This is another check in the safety review. To make sure your payload sails through the safety review, you should make sure it is spotless. Cleanliness levels for payloads waiting to be loaded on board a spacecraft can be found in NASA's *Contamination Control*

Requirements Manual, published in February 2012 by the agency's safety and Mission Assurance Directorate. Once you're sure your payload is surgically clean, you will need to demonstrate that it can safely contain any by-product of an experiment, liquid or solid, and also that there are no toxic materials that might be discharged into the cabin. It's probably not a good idea to fly a payload containing natural or manmade radioisotopes – in any quantity – either.

Emergency Decompression

Spaceflight, whether it's launching to the ISS or a suborbital jaunt, is a risky business and the possibility of an off-nominal event will be ever present, as evidenced by the tragic SpaceShipTwo disaster in October 2014. One such off-nominal event is rapid decompression. Since you and the pilot will be ensconced in your pressure suits, a rapid decompression event should be survivable, but what about your payload? To make sure your payload can survive a decompression event – explosive, rapid, or slow – you will need to test whatever you plan to fly in a hypobaric chamber. You will probably need to test in a freezer, because it gets pretty cold up there at 100 kilometers if there happens to be a puncture in the vehicle.

Electrical Power Interfaces

Chances are your payload will need electrical power so you will need to determine what your baseline energy allocation is to minimize power requirements for the simple reason that payloads with high power requirements will reduce the chances of being manifested. To determine this, you will need to consult with XCOR who can tell you what the total power and maximum continuous power available to your payload will be. To ensure continuous power is fed to your payload throughout the flight, you may need to fit an overload protection circuit (provided by a 10-amp circuit breaker). Another item on your electrical power checklist will be to ensure your electrical wiring insulation is rated to XCOR's requirements. Again, your PIM can check this for you. Bear in mind that, when the Lynx is on ground power, your payload may experience voltage transients and power ripples, so it's best to check what the power environment is when on the ground.

Electromagnetic Compatibility

If your payload produces an electromagnetic environment, you will need to utilize shielding to make sure your payload doesn't interfere with the operation of the Lynx. Equally, the vehicle you will be flying in will produce an interference environment with radiated electrical fields and it will be important that this radiated interference doesn't affect your payload.

OVERVIEW

There are a number of other checks that may be required – electrical bonding of equipment, computer interfaces, communication cables, wire harness shielding, to name just a few – but the purpose of this chapter has been to provide an introduction to the types of missions and the steps required to get your payload/science experiment accepted by XCOR. As I said earlier, you can make your life a whole lot easier by handing over all this payload integration work to a PIM. Good luck!

7

Passenger Training and Certification

Credit: XCOR

How much training will the new breed of commercial suborbital astronaut require? I think it is safe to presume he or she will require more training than the few days set aside by XCOR for its spaceflight participants and certainly much less training than that required by government trained astronauts preparing for increments on board the International Space Station (ISS). But where does the sweet spot lie? How long will it take to train a payload specialist to work efficiently and productively in an environment that will be extremely unforgiving of real-time snafus? And let's not forget what the demands of that environment are. First there is the wow factor brought on by those jaw-dropping views (Figure 7.1) and then there is the cost of the time spent in microgravity: \$150,000 divided by four minutes equals US\$37,500 per minute, or close to US\$625 per second. Don't drop

E. Seedhouse, *XCOR, Developing the Next Generation Spaceplane*,
Springer Praxis Books, DOI 10.1007/978-3-319-26112-6_7

7.1 The view from orbit. Credit: NASA

anything!! My suggestion to you if you plan to fly as a payload specialist is to gain as much experience in analog environments as possible. Become a scuba-diver, learn to fly an aircraft, and gain as much exposure to weightlessness on board parabolic flights as possible. You may be thinking why a payload specialist flying in the right seat of the Lynx would need all this supplementary training given that the astronaut will be strapped into their seat for the duration of their flight, but spaceflight – whether you happen to be strapped in or floating freely – is a profoundly disorienting experience, so it makes sense to spend as much time exposed to similarly challenging environments. In terms of what phases of training will be required, payload specialists will need to complete XCOR's vehicle familiarity training, which is included in the ticket price. That phase will take about three days. Beyond that, the payload specialist will need to complete training specific to operating their payload in addition to familiarization on board a parabolic flight. Total time? It depends on the complexity of the payload and/or science being flown, but training will take at least a week and perhaps as long as three.This chapter is all about teaching the average person in the street how to fly in space. Some will be buying a ticket on the Lynx for the thrill of rocketing into space, some will be on a mission to test a payload, and some will be scientists. But, no matter what their background, they will all need to be trained because once the Lynx starts flying we will be leaving behind the era of governments selecting astronauts based on intelligence and aptitude for spaceflight and we will be entering a period when astronauts select themselves based mostly on the thickness of their wallets. Way back when the Shuttle was flying, astronauts trained for at least a year for a two-week flight. Today, in the ISS era, astronauts typically train for four years for an increment lasting up to six months. Four years! With the advent of suborbital flights, we will have an extreme end of the astronaut training spectrum, with spaceflight participants requiring perhaps as little as three days of training. What will that training include? Well, below is a generic schedule of the sort of training XCOR will be delivering for its passengers.

Day One AM

Classroom:

Regions of the atmosphere
Altitude physiology and the hypobaric chamber
Unusual attitude flight profiles and Mach flight

Day One PM

Chamber:

Rapid and slow decompression in the hypobaric chamber

Classroom:

Acceleration physiology and the anti-G straining maneuver (AGSM)
Spacecraft safety and emergency egress
Lynx indoctrination: safety systems and mission architecture

Day Two AM

Classroom:

AGSM review, theory, and practical. Centrifuge manifest assignment

Centrifuge:

Gradual onset runs (GOR) # 1 and 2 (familiarization to 6 Gs)
Rapid onset runs (ROR) # 1 to 3 (ROR4 for 15, ROR5 for 15, and ROR6 for 15)
Debrief/review of G-videos

Day Two PM

Classroom:

Lynx life-support systems
Final frontier design (FFD) spacesuit indoctrination
Spacesuit donning and doffing (practical)
Spacesuit ingress and egress (simulator)

Chamber:

Armstrong line chamber run to 24,380 meters wearing spacesuit

Day Three AM

Classroom:

Flight briefing
Unusual attitude and high-G flight in Extra 300. 45 minutes

SPACESUIT

In 2008, Nassim Nicholas Taleb published *The Black Swan: The Impact of the Highly Improbable* and so the "black swan" event was born. A black swan event is one that is rare and difficult to predict, such as the 1987 stock market crash, the 11 September attacks, or the SpaceShipTwo accident (Figure 7.2). As I'm sure you're aware, SpaceShipTwo crashed in October 2014 when its feathering system deployed prematurely when the vehicle was traveling at Mach 1 at an altitude of 15,000 meters. Extreme aerodynamic forces caused the vehicle to disintegrate and the cabin suffered an explosive decompression. While the pilot Peter Siebold and co-pilot Michael Alsbury were wearing flight helmets and were hooked up to oxygen masks, they were not wearing pressure suits. The consequences were dire. Immediately, Siebold and Alsbury were exposed to −57°C and an atmospheric pressure of 15% of sea level, which meant they had less than 15 seconds of useful consciousness. Fortunately, even though Siebold was unconscious, his parachute opened automatically and he survived. Alsbury wasn't as fortunate.Ever since the X-Prize-winning flight of SpaceShipOne in October 2004, the Virgin Galactic pilots have worn one-piece flight suits because the thinking was that the pressurized cabin would be sufficient protection against the elements. Should they have worn pressure suits? After all, the U-2 pilots have been wearing them for decades, and this legendary reconnaissance aircraft routinely

7.2 The SpaceShipTwo accident. Credit: NTSB

7.3 U-2 pilot. Credit: USAF

operates at altitudes similar to those encountered by SpaceShipTwo during its testing program. With an operational ceiling of 21,000 meters (the actual ceiling is classified), the U-2 is used for weather surveillance and signal intelligence among other activities. Its pilots, who are attached to the 9th Reconnaissance Wing at Beale Air Force Base in California, have always worn pressure suits, provided by the David Clark Company, which also happens to be the same company that made the pressure suit used by Felix Baumgartner for his free-fall jump from 39,000 meters. These state-of-the-art suits (Figure 7.3) provide the pilots with oxygen and also ensure comfort by regulating temperature and humidity. In the event that a flight goes pear-shaped, the suits are more than capable of protecting a pilot from bail-out, even if the bail-out altitude happens to be above 20,000 meters.

Another aircraft that required its crew to wear pressure suits was the X-15, the hypersonic research vehicle that flew between 1959 and 1968. Built by North American Aviation and carried aloft to 13,500 meters by a B-52, the X-15 (Figure 7.4) featured a cockpit that became pressurized at 10,700 meters and pilots wore a pressure suit that supplied oxygen. In addition to the 13 test flights that exceeded 82,000 meters, the X-15 was flown into space on two occasions, each time piloted by Joe Walker in 1963. In addition to Walker's suborbital excursions, it's worth remembering the two suborbital flights of the Mercury

7.4 X-15. Credit: NASA

program: the first, piloted by Alan Shepard on 5 May 1961, reached an altitude of 187 kilometers, and the second, piloted by Gus Grissom, reached an altitude of 190 kilometers. Grissom and Shepard (Figure 7.5) wore pressure suits.And then there was the most seriously badass aircraft ever to take to the skies: the Blackbird, aka the SR-71. Back in the days of the Cold War, if the US wanted to know what was going on behind the Iron Curtain, it had to deploy an aircraft capable of flying very, *very* fast and very, *very* high. That aircraft was the SR-71 (Figure 7.6) and its elite pilots wore the most cutting-edge pressure suits available, which happened to be the David Clark Company Model 1030 back in those days. The organization responsible for fitting and testing the suits was the Physiological Support Division (PSD) at Beale Air Force Base facility that also housed an altitude chamber capable of testing to 26,000 meters. The 1030 suit comprised an inner rubber layer – the bladder – that was lined with tubes connected to cooling air, and between the bladder and the outer layer was a mesh material that prevented the bladder from over-inflating. As a confidence test, each SR-71 pilot was subjected to a rapid-decompression (RD) test to 26,000 meters. After struggling into the 1030 suit, they were strapped into the SR-71's ejection seat (wearing the pressure suit meant the pilots couldn't do this themselves) by PSD technicians inside the chamber. The chamber's door was closed and the decompression test began. Once 7,620 meters had been reached, the pilot was asked

7.5 Al Shepard. Credit: NASA

whether he had any sinus issues. If everything was clear, the ascent continued onward and upward to 26,000 meters as the pilot kept a close eye on a glass of water inside the chamber. As the altitude approached 19,200 meters (the Armstrong Line), it slowly began to boil, which was a stark reminder of what would happen to the pilot's body fluids if he wasn't wearing a pressure suit.After the 26,000-meter ceiling was reached, the altitude was brought down to 8,000 meters, which was the SR-71's cabin altitude. In an adjacent chamber, the altitude was brought up to 26,000 meters and the pilot was warned of the impending RD event. At the flick of a switch, the cabin altitude instantly rocketed up 26,000 meters accompanied by a loud bang and fogging. As a result of the rapid pressure change, the suits would become rigid to give pilots an idea of how difficult routine cockpit tasks – such as pulling the ejection handle, for example – would be following an RD.

> "We have always been clear that a shirt-sleeve environment was part of the baseline design. However, safety remains the priority, and should any new factors emerge that mean we should change that or any other element, then of course we will do so."
>
> *Virgin's Stephen Attenborough in a 2013 magazine article*

7.6 SR-71. Credit: USAF

"We think the safest thing is to not have people in pressure suits but to have them in flight suits and then in a cabin which protects them and allows them the freedom of microgravity because these people will be able to get out of their seats and float around the cabin."

Virgin Galactic's president and CEO George Whitesides, The Telegraph, *two weeks before the SpaceShipTwo crash*

Virgin Galactic's position on pressure suits seems to be contradictory given the history of manned suborbital spaceflight, doesn't it? After all, NASA and the USAF never for one moment considered flying their vehicles without their pilots wearing a pressure suit, so why did Virgin Galactic think it would be safe? Perhaps they based their mode of operations on the Concorde experience? Concorde was a supersonic passenger jet that was flown between 1976 and 2003 at an altitude of 18,000 meters. At this altitude, a puncture in the skin of the aircraft would have meant passengers would have had about 15 seconds of useful consciousness unless they managed to grab their oxygen masks. Another factor that may have swayed Virgin Galactic away from the requirement to wear pressure suits was the work attire of astronauts on board the ISS. Those working on the orbiting outpost wear nothing more than shirtsleeves and pants – an image that reinforces the perception that flying to space is safe. Of course, it is anything but, because astronauts are among the most highly trained humans on and off the planet, and there are myriad emergency protocols in place to protect crewmembers in the event of an off-nominal event. At this point, it is worth comparing the approaches of other vehicle providers to see what their perspective on pressure suits is. XCOR we know will require its pilot and passenger to wear a pressure suit, and those ferried on board Sierra Nevada's Dream Chaser will also be wearing spacesuits, so why does Virgin Galactic insist on only flight suits for its crew? After all, if a commercial passenger jet were to suffer a puncture, the oxygen masks drop down and the pilot dives to lower, breathable altitude. This will not be the case in a suborbital flight because, once the vehicle is on its way to space, it will be several minutes before it reaches a survivable altitude, by which time the passengers could be … well, let's not think about that.

> "What we do know is that even if Siebold did not experience ebullism, future space tourists – in the event of a cabin depressurization or spacecraft breakup – could. That's because the space industry has defined the 'outer edges of space' as 62 miles, or nearly 100 km, well past the Armstrong limit (the point at which water boils at 98.6 degrees Fahrenheit, the temperature of the human body). The possibility of ebullism (and other pressure-related elements) drives home the need for all passengers to don pressure spacesuits and oxygen masks, not T-shirts and shorts like some idealized visions of consumer space travel."
>
> *Michelle La Vone,* Space Safety Magazine, *December 2014*

Of course, Branson wants to make spaceflight fun and that will prove difficult if his passengers are tethered with oxygen hoses and constrained by bulky pressure suits but, in light of the SpaceShipTwo accident, the pressure suit issue may be one worth reconsidering.

ALTITUDE PHYSIOLOGY

One of the academic sections that will be delivered to prospective Lynx astronauts will be a lesson on altitude physiology. It's an important module because knowledge of basic altitude physiology helps you understand why pressure suits are so very, *very* important. But first some background. To earn your bragging rights as an astronaut, you need to fly to 100 kilometers of altitude, which is the internationally recognized boundary of space – unless you happen to be a Virgin Galactic passenger, in which case the altitude is a little lower.[1]

But, before we talk about what happens to the body in space, we need to understand what happens on Earth. We'll begin with the atmosphere, which is divided into layers (Figure 7.7). The lowest layer, which is the troposphere, is the one that we're most concerned with during most day-to-day activities. This layer also happens to be the most complex because there are so many variables that can affect conditions in this section of the atmosphere. Beginning at sea level and extending to an altitude of 7,000 meters at the

---

[1] X-15 pilots were awarded astronaut wings for flying above 80 kilometers, but the Fédération Aéronautique Internationale, FAI, the world governing body for astronautics records, defines space as an altitude above 100 kilometers. For those flying on Virgin Galactic, your ticket guarantees an altitude of 80 kilometers, which is not space, although Virgin Galactic passengers will be awarded astronaut wings – not FAI-branded, but Virgin Galactic-branded. The same applies to Lynx passengers incidentally: only the pilot will receive FAI/FAA-branded wings.

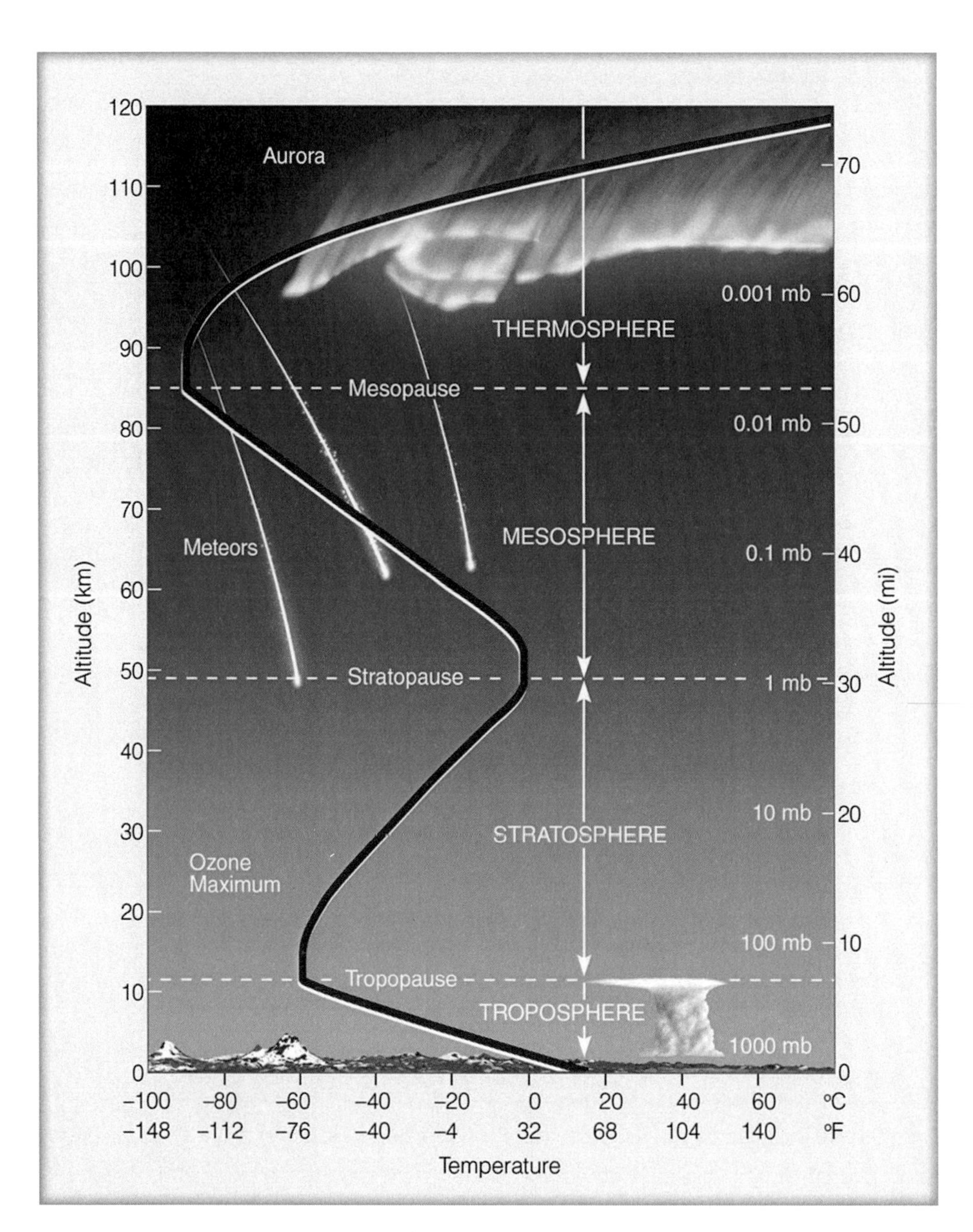

7.7 Layers of the atmosphere. Credit: NASA

7.8 WorldView. Credit: WorldView

poles and up to 18,000 meters at the equator, the troposphere contains about 75% of Earth's atmosphere. In this layer, temperature rises by six degrees Celsius for each 1,000-meter rise in altitude and, as you gain altitude, pressure also falls. At sea level, this pressure is 760 mmHg, but this number falls as altitude is gained, and this is when problems may be encountered.

Continuing up through the troposphere, we eventually encounter the tropopause – a boundary layer characterized by stable temperatures. Next is the stratosphere – a layer which will be familiar to all of those who followed Felix Baumgartner's Project Stratos. The stratosphere, which extends all the way up to 50 kilometers, contains the ozone layer and is defined by rising temperatures with increasing altitude – a characteristic that results in very stable atmospheric conditions. If you are a WorldView (Figure 7.8) passenger, then this is the layer you will be visiting in your gondola.

Stacked above the stratosphere is the mesosphere, which extends almost all the way to the top of our atmosphere at an altitude between 80 and 85 kilometers. In common with the trend in the troposphere, temperature falls with increasing altitude, but temperatures in the mesosphere are much colder than the lower layers and can be as low as −100°C: that's cold enough to freeze water vapor into ice clouds, which are known as noctilucent clouds – the subject of Project PoSSUM. Separated from the mesosphere by the mesopause is the thermosphere, which is the outer layer of our atmosphere. This layer extends all the way to an altitude of 640 kilometers and is marked by temperatures rising to as high as 1,000°C. Now that sounds hot, but you have to remember that up at thermosphere altitudes, there are so few molecules that there would not be enough energy for us to feel that heat. Beyond the thermosphere is the exosphere, which gradually merges into deep space.

Table 7.1 Time of useful consciousness

Altitude flight level	*Altitude (feet)*	*Altitude (meters)*	*Time of useful consciousness*
150	15,000	4,572	30 min or more
180	18,000	5,486	20–30 min
220	22,000	6,705	5–10 min
250	25,000	7,620	3–6 min
280	28,000	8,534	2.5–3 min
300	30,000	9,144	1–3 min
350	35,000	10,668	30–60 sec
400	40,000	12,192	15–20 sec
430	43,000	13,106	9–15 sec
500 and above	50,000	15,240	6–9 sec

So that's our atmosphere. But what happens as we fly through those layers? Well, the most important point to remember is that as altitude increases so does air pressure, which is why spacecraft cabins are pressurized. If you have flown commercial, you may have noticed a faint hissing sound once the doors have closed. That's the sound of the aircraft being pressurized. When your aircraft takes off, the cabin pressure is slightly higher than the outside air pressure and, as the aircraft ascends, the pressure is adjusted to decrease the differential between the internal pressure and outside air pressure, which is why most commercial aircraft fly with a cabin pressure of around 1,600 meters. If the cabin pressure was a little higher (say 2,500 meters), people with respiratory problems would have difficulty breathing and, if the cabin pressure was a lot higher (say 4,000 meters), then all the passengers would experience low-pressure symptoms – headaches, dizziness, vomiting – and that would be bad for business. Reduce the altitude some more and eventually all the passengers would lose consciousness (Table 7.1).

The reason we humans lose consciousness at high altitudes is because our brain cells just weren't designed to function properly at low pressure, hence the need for those bulky pressure suits we talked about earlier (see sidebar). Here's some history. There are two primary types of pressure suit: the full pressure suit and the partial pressure suit, although this latter term is a little misleading because both suits fully protect the human inside the suit, it's just that a different approach is used depending on the suit. The full pressure version (EVA) (Figure 7.9) encloses the human completely in an airtight pressurized suit, whereas the partial pressure version (Figure 7.10) relies on figure-hugging material and tubular air compartments running parallel to the arms to exert pressure on the wearer's skin.

No Pressure Suit?

Those of you who have watched *2001 A Space Odyssey* will no doubt remember the explosive decompression scene. In this scene, Bowman (Keir Dullea) is performing a spacewalk inside an escape pod and is prevented from re-entering the *Discovery* by the HAL-9000 computer. But Bowman blows the bolts on the pod and enters *Discovery*'s airlock – an act that causes him to be exposed to a vacuum for 14 seconds before he can re-pressurize the compartment. Fiction or fact? Actually, such an exercise would be survivable, although not advisable. Way back in the 1960s, the USAF subjected chimpanzees to explosive decompression which left the hapless test subjects exposed to vacuum for more than three minutes in some cases. All of the guinea pigs survived except one. If you were to be exposed to an explosive decompression event, then we know you would have up to 10 seconds to help yourself, so think quickly! After 12 or 13 seconds you would begin to experience impairment and, if you weren't wearing a pressure suit, your body would begin to swell because the liquid in your soft tissues would begin to vaporize. Contrary to what you may have seen in Hollywood movies – the decompression event in *Outland* comes to mind – you would not explode because the skin is extraordinarily resilient. After that swelling sensation is noticed, your blood would stop circulating. At this point, you will have been exposed to a vacuum for around 60 seconds. Next, the vacuum would go to work on your lungs with catastrophic results because your pulmonary system is perhaps the most vulnerable to extreme low pressure. So, if you are exposed to a vacuum due to an explosive decompression, there is time to take action. But be quick!

7.9 EVA. Credit: NASA

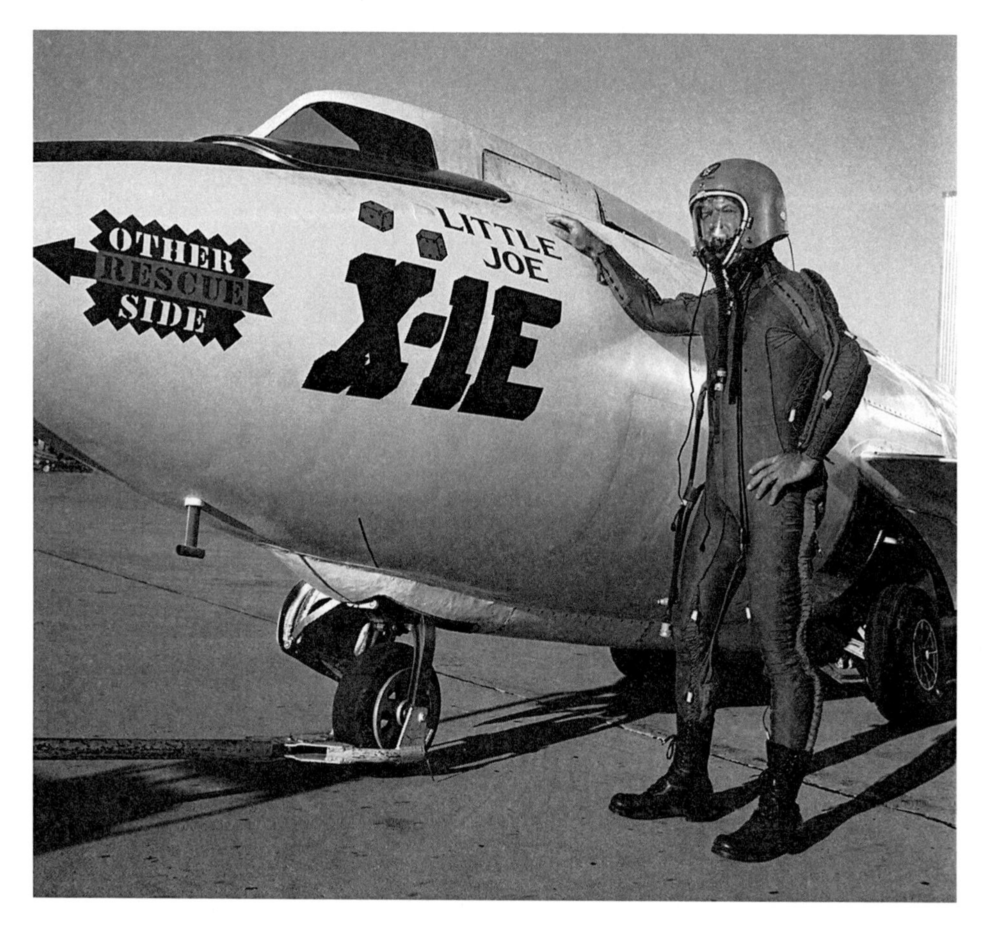

7.10 X-1. Credit: NASA

Full pressure suits, which were descendants of undersea diving suits, were the first to be tested for high-altitude operations by scientists in the US and Great Britain in the 1930s, whereas the partial pressure suit was born out of a need for such a garment in the Second World War. Much of the research conducted on partial pressure suit design was carried out by Dr. James Henry at the Biophysics Branch of the Army Air Force's Aero Medical Laboratory, which today is known as Wright-Patterson Air Force Base. Much of Henry's work on the design on the partial pressure suit came from the world of G-suit protection for fighter pilots, which is why his mechanical pressure suit featured tubular assemblies (capstans – in the G-suit, these devices were inflated to maintain blood pressure). Henry's partial pressure suit was capable of protecting a pilot at altitudes as high as 25,000 meters, which was a performance capability that was so impressive that the army's high command recommended that further research be carried out in cooperation with the David Clark Company which was a prime contractor for the US Army's G-suits at the time. During the post-war cooperative research, Dr. Henry and the David Clark Company developed the S-1 and S-2 suits, which, after a series of refinements, morphed into the MC-series suits in the

1950s and 1960s. But, by the 1960s, interest in the partial pressure suit was waning due to an emphasis on full pressure suits that were required for those crewmembers involved in projects such as the X-15, X-20 (DynaSOAR), the Manned Orbital Lab, and Project Mercury. These space and near-space projects demanded very high-altitude life-support requirements which could only be met by the full pressure suit and so the partial pressure suit died a natural death, with the exception of a space activity suit that was developed as part of a special NASA program in 1971.

Today's full pressure suits are a layered system that act together to protect the occupant. At its simplest, the suit is a balloon filled with air and the balloon in this case is the bladder, which acts as a restraint layer that prevents the suit from rupturing. Other restraint layers may be included, such as anti-G layers that help counteract the effects of G during high-G turns or during launch and re-entry. The exterior layer, which is usually brightly colored, is often made out of fire-retardant material (Nomex). For many decades, the basic design of the full pressure suit has remained the same, but changes may be on the way thanks to Dr. Dava Newman, a MIT bio-astronautics engineer (and now Deputy Administrator of NASA) who developed the BioSuit (Figure 7.11).

Classed as mechanical counter-pressure (MCP), the BioSuit appears more like a skin-suit than a standard full pressure suit, but it is still designed to do the same job. The

7.11 Dava Newman wearing the revolutionary BioSuit. Image courtesy: Professor Dava Newman, MIT: Inventor, Science and Engineering; Guillermo Trotti, A.I.A., Trotti & Associates, Inc. (Cambridge, MA): Design; Dainese (Vicenza, Italy): Fabrication; Douglas Sonders

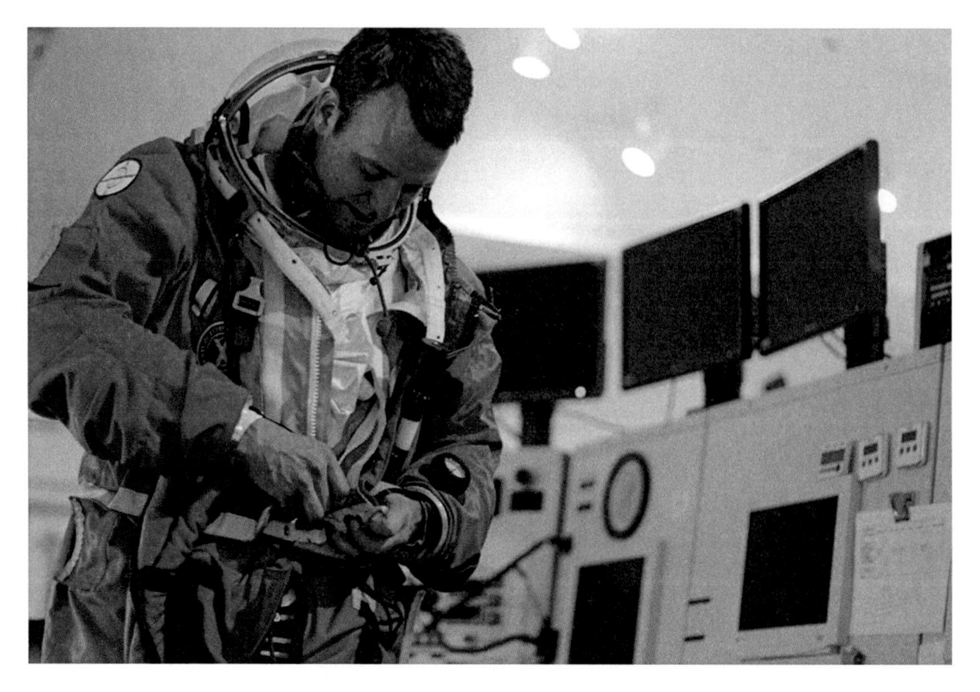

7.12 Ted Southern. Credit: Final Frontier Design

principle behind the suit is applying the same pressurization as is applied by a standard full pressure suit but applying that pressure directly to the skin in a way that sidesteps the need for gas pressure. Newman and her team reckon this can be achieved by using the latest in active compression fabrics. If they are right, then future astronauts will be able to do away with the bulky restrictive balloon suits and look forward to a day when they can don a lightweight high-mobility suit instead. But what about the suit you will be wearing in the Lynx? Well, there is a good chance it will be provided by the innovative engineers at FFD, a spacesuit company based in Brooklyn. FFD came about as a result of a meeting between Ted Southern and Nik Moiseev in 2007. Nik had spent the best part of 20 plus years working in the Soviet Union designing cutting-edge spacesuits, whereas Ted was an artist who had a hand in designing Victoria Secret garments (the angel wings happen to be one of Ted's signature items). As competitors in NASA's Centennial Challenge to design an astronaut glove, Ted and Nik's design was placed second, which was good enough for US$100,000 in grant money from the agency. FFD was born out of that grant money and the rest is history. Today FFD is at the cutting edge of spacesuit design (Figure 7.12) and the company is already working on their third-generation suit which has been tested by Project PoSSUM astronauts in Embry Riddle's suborbital simulator.

HIGH-ALTITUDE INDOCTRINATION (HAI) TRAINING

To train you how to react to an RD, operators will require you to complete high-altitude indoctrination (HAI) training. This will begin with a ground school component which will cover the basic principles of altitude, which we have already covered in this chapter.

7.13 Hypobaric chamber. Hospital Corpsmen 2nd Class Kyle Carswell and Daniel Young monitor members of the 2009 class of NASA astronaut candidates for hypoxia in an altitude chamber. Credit: NASA/USN (File: US Navy 091006-N-9001B-017)

You will then be introduced to the altitude chamber (Figure 7.13) where the next phase of training will begin (see sidebar). Here you will be assigned seats, given a brief overview of the flights and what you can expect at each flight level, and a safety briefing. You will be accompanied in the chamber by one or two observers who will talk to the physiological training officer or flight surgeon observing the flight from the console outside the chamber.

The MACC Suit Chamber

Capable of flying to 30,480 meters of altitude and supporting RD rates of sea level to 30,480 meters in less than five seconds, the MACC Suit Chamber comprises a two-meter-diameter cylinder that seats two fully suited astronauts. It's the perfect environment for all manner of astronaut training, including simulating flight profiles, practicing emergency procedures, evaluating suit performance in low-pressure conditions, and building confidence while wearing a pressure suit during rapid or slow decompression.

When it is time for your flight, you will find the chamber crew ready and waiting with the chamber prepped and ready to go. You will be shown into the chamber, don your helmet and mask, hook up your mask, and check your communications and oxygen systems. The inside observers will then complete a physical check of everyone's oxygen connection to make sure the connections are secure. This will be followed by a communication check beginning with the students and ending with the chamber crew. As chamber technicians ready the chamber for the flight, the inside observers will go through some of the material covered in the lectures by asking the students questions related to altitude physiology and safety procedures during the flight. With the final checks complete, the hatch will be closed and the flight director will give a thumbs-up, indicating that the flight is about to begin. Once the flight director receives a thumbs-up from each student, the chamber will begin its ascent to 1,500 meters at a rate of 1,500 per minute. At 1,500 meters, the chamber will level off and the inside observers will ask each student to complete an ear and sinus check followed by a confirmatory thumbs-up. If everyone has clear sinuses, the chamber will return to sea level and the inside observers will once again ask for a thumbs-up to make sure everyone is ready for the hypoxia demonstration flight to 7,620 meters. After leveling off at 7,620 meters, the students will be divided into two groups. The first group will drop their oxygen masks to experience hypoxia after being told to put on their oxygen masks as soon as they experience one clear-cut symptom of hypoxia. As a flight director, I have been in charge of dozens and dozens of chamber runs and it's always interesting to see how people react. Pilots tend to be the worst when reporting symptoms because this group is a competitive lot and each pilot wants to have the bragging rights of having been without oxygen the longest, or at least for the six minutes permitted. While off oxygen, the students perform simple tasks such as drawing cats and dogs, subtraction and multiplication, and answering general-knowledge questions. These tasks are intended to demonstrate to the students just how insidious hypoxia can be. During the exercise, the students are confident that they are completing the tasks correctly, but the results – which are presented to the students following the chamber run – often tell a different story. To help the inside observers monitor symptoms, the group experiencing hypoxia wear pulse oximeters that display oxygen saturation. These devices don't lie, although this doesn't stop some trying to push through their hypoxia symptoms. I remember one fighter pilot who had been off oxygen for more than five minutes who insisted he had no hypoxia symptoms despite his pulse oximeter reading 61%! On another occasion we had a guy who completed the six minutes off oxygen with no clear hypoxia symptom. After the chamber run, we asked him some background questions which revealed he was a 40-a-day smoker: this guy's body was in a permanently hypoxic state! Every once in a while, a student pushes the hypoxic limit and it is up to their buddy, sitting opposite wearing their oxygen mask, or the inside observer to step in and hook up the incapacitated student to their oxygen mask. Once everyone in the first team has experienced one clear-cut symptom of hypoxia and is hooked up to their masks, it is the turn of the second group to drop their masks and the exercise is repeated. After the flight, the students are quizzed about their hypoxia symptoms (Table 7.2). Some lose their color vision, some become dizzy, and others feel tingling in their extremities. One interesting characteristic of hypoxia is that symptoms can be very variable. Just because you experienced a particular symptom one day doesn't mean you will experience the same symptom the next time, so it's helpful to know the range of symptoms.

Table 7.2 Hypoxia symptoms

Stages	*Indifferent: 90–98% oxygen saturation*	*Compensatory: 80–89% oxygen saturation*	*Disturbance: 70–79% oxygen saturation*	*Critical: 60–69% oxygen saturation*
Altitude (1,000 feet)	0–10	10–15	15–20	20–25
Symptoms	Decrease in night vision	Drowsiness Poor judgment Impaired coordination Impaired efficiency	Impaired handwriting Impaired speech Decreased coordination Impaired vision Impaired cognitive function Impaired judgment	Circulatory failure CNS failure Convulsions Cardiovascular collapse Death

The day after their hypoxia demonstration, the students enter the chamber for a second flight to 13,000 meters. The purpose of this flight is to demonstrate positive pressure breathing (Figure 7.14) and also to provide another opportunity for students to practice clearing their ears during the descent. At 3,000 meters, students are told to drop their masks, although the inside observers keep theirs on until the chamber reaches ground level. Two chamber runs down, two to go. The third chamber run is the RD flight, which is intended to simulate an immediate loss of pressure as a result of a puncture in the skin of the Lynx and spacesuit.

Before describing what happens during the RD, it is helpful to understand the layout of the chamber. The chamber has two sections, one of which is the main chamber and the other the secondary chamber or lock. The chambers can operate independently thanks to a hatch and valve separating the two, which means that the main chamber can run at a different pressure than the lock. During the RD run, the main chamber is sealed and flown to 12,000 meters. As the main chamber is being flown, the inside observer and the group of students enter the lock, complete their hook-up and checks, and fly to 2,400 meters. As they ascend, the inside observer reminds the students what they can expect during the RD and, once the lock reaches 2,400 meters, the chamber engineer prepares to open the valve separating the two chambers. Once the inside observer has briefed the students to expect an RD, all the students can do is wait. But not for long. After a few seconds, the chamber engineer pushes the button opening the valve. A moment later, there is a loud bang that gets everyone's attention, followed by fogging and a rush of wind as the air in the lock is rapidly evacuated. It's always interesting to watch how students react to an RD. Even though everyone knows an RD is imminent, that bang is loud enough that everyone jumps as if someone has just stuck a wide-bore needle into them. Some sit there with a "deer in the headlights" look and have to be prompted to carry out their safety checks, but most recover from the shock and busy themselves with checking their connections as instructed before giving the inside observer the required thumbs-up. With the RD over, all that remains is the final and fourth flight to 24,000 meters (see sidebar) (Figure 7.15).

7.14 Fighter pilot. Credit: USAF

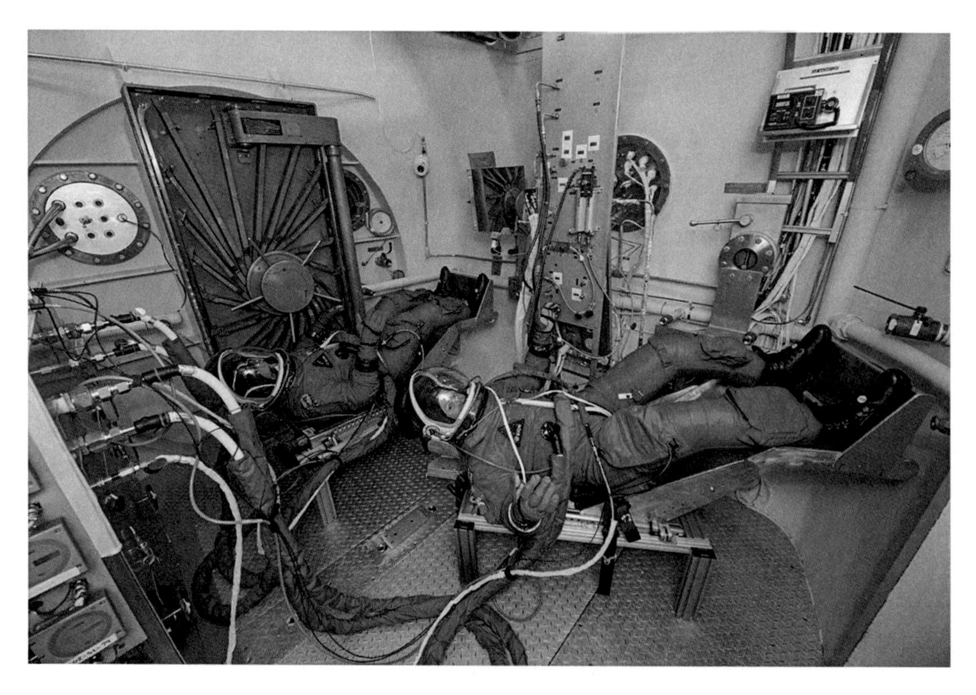

7.15 Pressure suit test. Credit: NASA

The MACC Chamber

The MACC chamber located onsite at Midland Space Port is capable of ascending to 30,480 meters and can accommodate up to 10 astronauts wearing their spacesuits. It's a very spacious chamber that has enough room to fit an entire vehicle cabin, thereby providing astronauts with a high-fidelity and very realistic training platform. Seated in their mock-up cabin, astronauts can not only test their spacesuits, but simultaneously check interfaces within the cabin, assess mission performance, practice manual operations, learn how to react to emergencies, and simulate an entire spaceflight from launch to landing.

One of the most important objectives of all this chamber training is to expose suborbital astronauts to simulated altitude so they can learn about their limitations and dangers of working in what is a very, *very* dangerous environment. It also provides an ideal opportunity for this group of astronauts to familiarize themselves with the spacesuit they will be wearing on their flight, hence the flight to 24,000 meters. After donning the spacesuit with the assistance of two spacesuit technicians, the astronauts will ascend to 24,000 meters, keeping a close eye on that glass of water that was mentioned earlier while simultaneously listening to instructions from the flight director. As the chamber ascends to altitude, the astronaut will be told how the suit should feel as the suit inflates. As the chamber passes 19,000 meters, the astronaut will notice that the suede patches will begin to smoke and the flight director might mention that, if the astronaut wasn't wearing a suit at this altitude, they would be dead. At 24,000 meters, it is time to perform some simple tasks such as pulling a pen from a pocket, checking D rings, finding the drink bottle, performing a simple press to test, and going through the emergency checklist. Once the astronaut has completed the tasks, he or she will be left at altitude for a few minutes so they can fully appreciate the potentially lethal situation they are in and also to build confidence in the suit. With the familiarization to 24,000 meters over, the chamber will descend to 7,620 meters and level off in preparation for the final chamber test: with the punch of a button on the chamber console, a loud bang, and a tremendous rush of air, the two chambers will equalize at around 21,000 meters. After their second RD in as many days, the astronauts will descend at 1,500 meters per minute to ground level, they will doff their suits, post-flight records will be written, and a briefing will be conducted. Then it will be time for centrifuge training.

ACCELERATION PHYSIOLOGY

In addition to spending a number of years as a flight director for the hypobaric chamber flights in the Canadian Forces, I was also lucky enough to work as Acceleration Training Officer, which meant strapping pilots into the centrifuge at Downsview (Figure 7.16). As with HAI, the practical element of acceleration training was preceded by a theoretical component on the subject of acceleration physiology. Let's begin with the basics and start with the unit of measurement: Gs. If you are a fan of Formula 1, you will have no doubt heard the

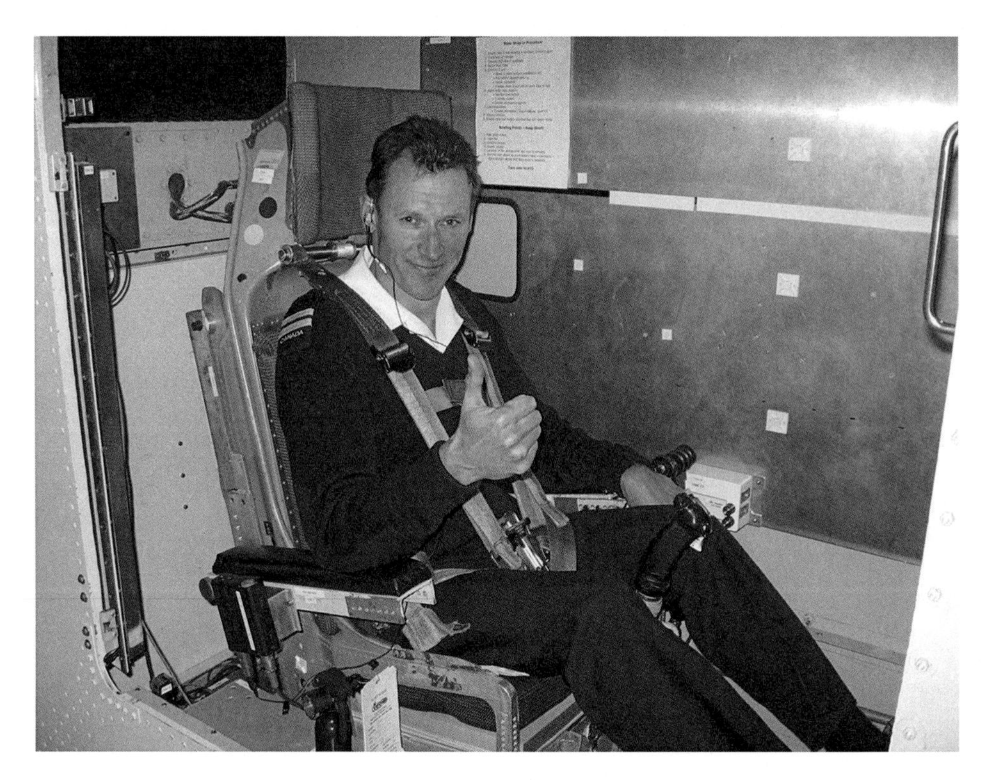

7.16 Author in the Downsview centrifuge. Credit: Author's collection

commentators talk about how many Gs the drivers are subjected to in the many high-speed corners that make up your standard Formula 1 track. In some of these corners, Lewis Hamilton, Sebastian Vettel, and Jenson Button may have to deal with lateral G-forces that exceed 5 Gs. When you consider that the average Formula 1 track has 16 or 17 corners, with perhaps half a dozen of those being high-G corners, and that your average Formula 1 race is more than 60 laps, you can begin to appreciate just how fit these guys are. But, if pulling 4 or 5 Gs repeatedly sounds like a workout, consider your average fighter pilot who may be subjected to more than 8 or 9 Gs. And these Gs may be sustained for several seconds. Whereas Fernando Alonso and his Formula 1 colleagues are subjected to Gs for a second or so, fighter pilots may have to deal with high-G loads for as long as five or six seconds. I remember a presentation given by the Flight Surgeon for the Blue Angels that included in cockpit footage of a turn in an F-22 that pegged the G-meter at 7 Gs for 22 seconds!

At this point in the introduction, it is important to note the different types of G, which are linear, radial, and angular (Figure 7.17). Linear acceleration is the sort of acceleration that is experienced during take-off or by Formula drivers at the start of a race, while radial acceleration is the type of acceleration a pilot is subjected to during a sharp turn or when pushing into and pulling out of a dive. The third type of acceleration is angular, which occurs during a simultaneous change in speed and direction, which happens during a spin or a climbing turn. The G-forces induced by these types of acceleration are abbreviated as Gx, Gy, and Gz (Figure 7.18).

7.17 Unusual attitude. Credit: USAF

Gx is the force that acts from chest to back and is experienced during take-off, Gy is the lateral force that is familiar to aerobatic pilots when they perform aileron turns, and Gz is the force that acts through the vertical axis of the body, from head to foot or from foot to head: if Gz is experienced from head to foot, it is termed positive Gz (+Gz) and, if acceleration is transmitted from foot to head, it is termed negative Gz (−Gz). As you can imagine, all these G-forces exert a significant strain on the body, particularly the cardiovascular system, which must keep blood flowing to the brain. While the cardiovascular system responds quickly to increased acceleration by increasing the heart rate, there is a point at which the physiological responses cannot keep pace with the Gs. When that happens, the cardiovascular system cannot pump sufficient blood to the brain and pilot performance is degraded, sometimes with fatal consequences. One of the first signs that things are going pear-shaped is loss of vision (LoV) because the eyes are particularly sensitive to low blood flow. As the Gs pile on, vision becomes more and more compromised and the pilot may suffer tunnel vision as his or her peripheral vision may become degraded. If the onset of Gs continues, the next sign may be gun-barrel vision which will be followed in short succession by grayout and blackout. At the blackout phase, the pilot is still conscious but cannot see anything – a sign that G-induced loss of consciousness (G-LOC) is imminent.

A pilot suffering from G-LOC may be unconscious for up to 15 seconds and it may take another 15 seconds for the pilot to recover their bearings and regain control of the aircraft, at which point it may be much too late. Fortunately, the symptoms of high-G exposure are fairly predictable, and pilots are trained to recognize these. For example, when G onset is gradual (0.1 Gs per second), visual symptoms normally precede G-LOC whereas, if the onset is rapid (1 G per second or greater), then G-LOC can be almost instantaneous. What

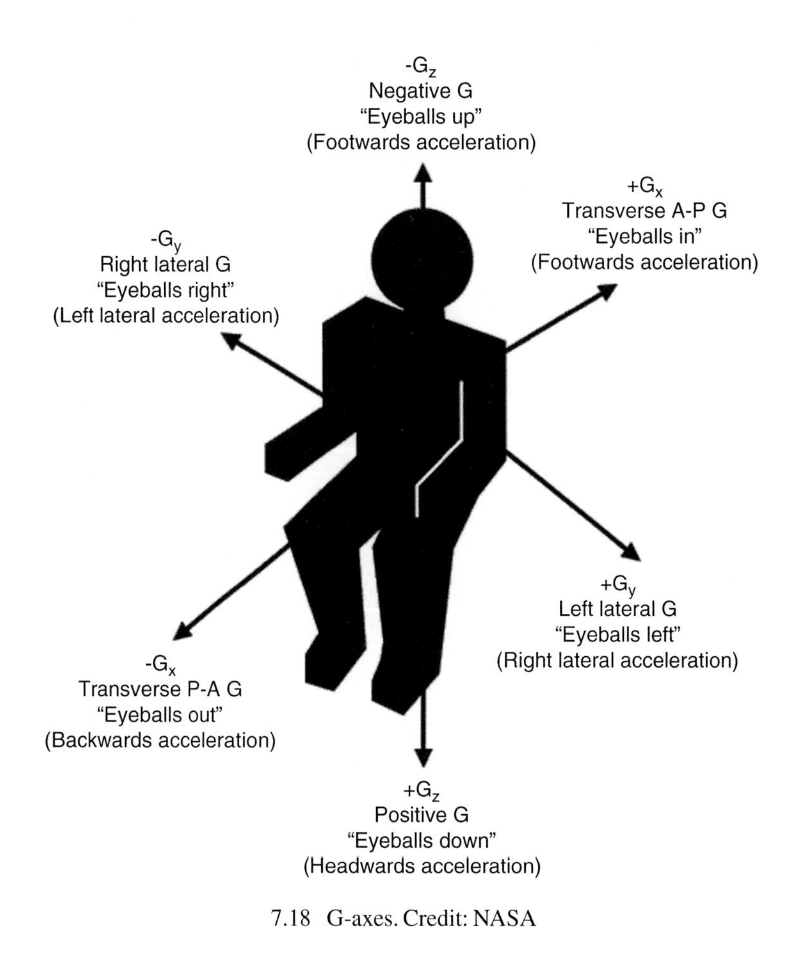

7.18 G-axes. Credit: NASA

does all this have to do with a suborbital astronaut flying on the Lynx? Well, under normal flight conditions, there is a low risk of either the pilot or his passenger suffering a G-LOC event but, if things go squirrely, then those Gs could pile on rapidly, and that is when all this acceleration training will be invaluable. So what can you do to increase your G-tolerance? Well, first of all, G-tolerance is degraded by alcohol, fatigue, and dehydration, so don't drink, get plenty of rest, and drink plenty of water before your flight. Second, practice your AGSM (see sidebar) on a daily basis in the two to three weeks leading up to your flight.

AGSM

The AGSM is a technique taught to all fighter pilots to increase their tolerance to the dreaded G. A perfectly executed AGSM will increase your tolerance by 2–3 Gs, so it's worth becoming proficient. If you are interested in seeing yours truly performing the AGSM while being spun in the centrifuge, you can enter my name in Google and add the term "centrifuge training." Enjoy!

The technique is all about timing and requires a deep inhalation of air while simultaneously tensing the big muscles in the legs, stomach, and buttocks. Following a count of three, the pilot exhales rapidly, inhales again, and repeats the exercise. In my role as Acceleration Training Officer, it was one of my responsibilities to ensure every pilot entering the centrifuge was capable of performing a proficient AGSM, which is why each pilot was required to demonstrate their technique to either myself or one of my instructors before entering the "fuge."

If you are sitting down reading this book, your blood pressure will likely be around 120 mmHg systolic and around 75 mmHg diastolic. Systolic pressure, which is the highest pressure, is attained as the left ventricle contracts and ejects blood into the aorta, and diastolic is the minimum pressure that is measured just prior to the next beat of the heart. Together, your systolic and diastolic pressure is a function of your heart rate and the peripheral resistance as your blood makes its way around your circulatory system. The reason this is relevant to acceleration is because a significant percentage of that blood flow must be channeled to your brain and that blood must be pumped uphill (in the standing or seated position, your brain is above your heart), which results in a loss of pressure. By the time arterial blood makes its way to your brain, there is an arterial pressure drop of around 35 mmHg and, if that arterial pressure drops some more, then there is a concomitant fall in pressure in the brain. Now imagine you are being accelerated forwards during your suborbital flight and the acceleration, or +Gz, is four times the normal acceleration of gravity. This will lead to an acceleration-induced pressure drop of 4 × 35 mmHg, or 140 mmHg, at the brain. As the Lynx rockets upwards and acceleration continues, your blood will flow "downhill" to your extremities, especially your stomach and legs. Your blood will pool there because the venous return of your heart will be compromised by that acceleration, which means the amount of blood being pumped by the heart is reduced and arterial pressure is further reduced. From this point on, if no fail-safe mechanisms are implemented, those 4 Gs of acceleration will become fatal as blood flow to the brain eventually spins down to zero. Fortunately, your body is equipped with some fail-safe mechanisms. One of the first things the body does when faced with acceleration is to increase heart rate. This compensatory mechanism acts in tandem with the pressure receptors that can be found at strategic locations in the circulatory system. These pressure receptors – baroreceptors – keep the brain informed of blood pressure and send signals to the brain whenever blood pressure levels are too high or too low. These compensatory mechanisms work well, but they are limited, and it isn't just the circulatory system that is affected. As you encounter those Gs during launch, your dense tissues will be driven downwards: for example, your liver will sink into your stomach and your heart will also sink into your chest, with the result that pressure will be exerted on your diaphragm. And as those Gs continue to pile on, your diaphragm will be displaced, which will make breathing increasingly difficult. As the G-meter scrolls past "3" you will feel a "dragging" sensation in your chest and stomach and as the Gs hit "4" you will be struggling for breath. Fortunately, thanks to your training in the centrifuge and by executing a proficient AGSM, you won't have any trouble tolerating the Gs – hopefully. I say hopefully because it is not by any means certain that everyone will be medically cleared for this challenge to the cardiovascular system and here's why. While the effects of +Gz and –Gz are well documented, there are certain effects of acceleration that don't appear on an electrocardiogram

(ECG). We know that heart rate increases and we know that vascular return is diminished under G and we know that all the muscle straining while performing the AGSM drives up systolic pressure. We also know that, if the AGSM isn't performed in synchrony with the G-loading, flow resistance has a tendency to fluctuate and that can be bad news for the heart. Why not perform the ECG while exercising, you may ask? The problem is that a treadmill exercise protocol will not cause a drop in cardiac output so a stress ECG won't tell you any more than a resting ECG will. In short, there is no way of assuring a potential suborbital astronaut that they can safely be exposed to sustained acceleration stress. If someone has a problem in the centrifuge, it is simply a case of punching a big red button on the control console, the fuge slowly grinds to a halt, and the rider can be escorted to the flight surgeon's office for a medical review. Not so en route to 100,000 meters! In fact, if a cardiovascular abnormality was to manifest itself in the early phase of a suborbital flight, that flight could potentially be life-threatening. Now, you may be thinking that the aeromedical community would have a good handle on this since thousands and thousands of pilots have been tested in centrifuges over the years, and you would be right. In a study conducted at the USAF School of Aerospace Medicine that examined 1,180 centrifuge training sessions, a whopping 47% resulted in arrhythmias and more than 4%of these should have resulted in termination of the run. The subjects of this study were aeromedical course students who had been medically pre-screened and were a healthy group. But, despite being screened, there was a significant number who had potentially harmful responses to acceleration. Another study that examined 195 fighter pilots revealed a rate of 2.6% ventricular tachycardia, 1.5% paroxysmal supraventricular tachycardia, and 0.5% paroxysmal atrial fibrillation. Each of these conditions is a red flag that would prevent someone from performing centrifuge training. So what to do? Well, I suggest going beyond the minimum flight medical and insisting on a comprehensive cardiovascular examination, especially if you are over 40 because those compensatory mechanisms discussed earlier tend to become less and less effective with increasing age. Another precaution you can take is to be instrumented during your ride in the centrifuge, which is discussed here.

The few centrifuges that exist worldwide vary widely in their capabilities. Some, like the Star City behemoth (Figure 7.19), have an onset rate exceeding 12 Gs per second while others, such as the one I was in charge of at Downsview, struggled to spin up at its advertised 3 Gs per second (it actually never managed more than 2.8 Gs per second). But it isn't just the onset rates that differ: some fuges, especially those focused on crew training, are fitted with closed-loop profiles and target tracking to make the whole acceleration experience as realistic as possible. Perhaps the gold standard in the centrifuge world is the Phoenix 4000 at NASTAR. Located in Southampton, Pennsylvania, this luxury fuge (Figure 7.20) is extremely versatile, which is one reason why it has been used by Virgin Galactic to train its spaceflight participants.

In standard aircrew training, pilots are required to complete a series of GOR, ROR, and high sustained G (HSG) runs: a GOR is defined as an onset rate of 0.1 Gs per second, a ROR is defined as an onset rate of at least 3 Gs per second, and an HSG run is run in which a pilot is subjected to 7 Gs for 15 seconds wearing a G-suit or 5 Gs for 15 seconds without G protection. Those are some tough runs! Fortunately, the Lynx won't be subjecting its passengers and crew to HSG but the acceleration forces will still be quite substantial, which is why centrifuge training will be required (see sidebar). The G-training course you will complete

7.19 Star City centrifuge. Credit: Harald Illig

7.20 NASTAR's centrifuge: the STS-400. "Flights" in the STS-400 are preceded by classroom lectures on the subjects of acceleration and the physical effects of spaceflight. Credit: NASTAR

7.21 Desdemona. Credit: AMST

has one primary and three supplementary objectives. The most important objective is to increase your G-tolerance by improving the effectiveness of your AGSM. The secondary objectives include providing soon-to-be suborbital astronauts with a better understanding of the physiological stresses of increased G, increased confidence in their ability to tolerate high Gs, and a better appreciation of the hazards encountered in a high-G environment. A G course typically begins with some theory in the form of a couple of lectures introducing you to some basic acceleration physiology and an overview of the runs. This is followed by some one-on-one AGSM training with one of the instructors, and then it's off to the fuge! The first run is usually a relaxed GOR that continues until peripheral light loss (PLL) is encountered; this is done to establish relaxed G-tolerance. The relaxed GOR is followed by a ROR specific to the vehicle, so you can expect a run that takes you up to 4 Gs.

Desdemona

Operated by TNO Netherlands Organisation for Applied Research, Desdemona (Figure 7.21) is a three-axis flight simulator mounted on the arm of a centrifuge. Its modular configuration means a cabin can be mounted on the arm which can then – thanks a fully gimbaled system – be rotated and/or spun in just about any axis. Perfect for training budding astronauts! This is probably why XCOR has been working with TNO to develop a simulated suborbital mission that is "flown" on

Desdemona. The fuge mission begins with a high-G boost from Spaceport Spaceport Curaçao followed by the microgravity phase during which the pilot and passenger can gaze down on a simulated view of the Caribbean from 100 kilometers of altitude. After a few minutes of simulated weightlessness, the nose pitches down, the re-entry Gs begin piling on, and the Lynx glides to a graceful landing. Desdemona is an amazing tool. So amazing that some pilots have pronounced the Desdemona suborbital simulation more stressful than the real thing.

SPACE MOTION SICKNESS

Nausea, vertigo, headaches, vomiting, and general discombobulation – these are all symptoms of space adaptation syndrome or space motion sickness (SMS), a syndrome that can strike just about anyone. Without the familiar pull of Earth's gravity, spacefarers face an environment that challenges the sensory system to its limits and, in a world where up and down are nowhere to be found, space travelers often find themselves the worse for wear. For as long as there have been astronauts, SMS has wreaked havoc for mission planners because, despite the best efforts of space life scientists over the years, nobody has a grip on the problem. Despite all sorts of medications having been used over the years, astronauts still find themselves disorientated and queasy during the first couple of days of a mission. Typically, more than half of first-time astronauts suffer from SMS and symptoms usually resolve within 72 hours. Second-time flyers suffer fewer symptoms and third-time flyers are practically symptom-free, but sending astronauts into space repeatedly so they can adjust to the disorienting environment is a very, very expensive way of dealing with what is an intractable problem. So, if you want to avoid being the crewmember that has to spend half their mission wiping the contents of their stomach off the console, what do you do?

Well, firstly it is helpful to understand the problem as it applies to suborbital flight. We'll begin with the hypotheses that have been put forward to explain why SMS occurs – the fluid shift hypothesis and the sensory conflict hypothesis. The first of these suggests SMS is caused by the fluid shift towards the head and chest region caused by the loss of hydrostatic pressure in the lower body when astronauts enter microgravity. This fluid shift, which can be as large as two liters, causes an increase in intracranial pressure (ICP) and an increase in cerebrospinal fluid (CSF) pressure, which exerts pressure on the vestibular system, thereby inducing SMS. The sensory conflict hypothesis on the other hand argues that SMS is caused by the conflict in vestibular and visual cues that occurs in microgravity (Figure 7.22). The end result of this fluid shift and/or sensory conflict is a range of symptoms (Table 7.3) that cause a major headache for an astronaut.

One serious consequence of an astronaut's sensory system being out of sorts is responding to an emergency, since, with the crewmember's perceptual-motor system compromised, the astronaut will find it difficult to perform tracking tasks, switch throws, and fine manipulation tasks. Given the negative operational consequences of SMS, it isn't surprising that space agencies have spent a lot of time and resources in developing countermeasures. The Russians had some success with pre-flight stimulation of the vestibular system and similar pre-flight behavior modification techniques, which spurred an interest on the application of

Table 7.3 Symptoms of motion sickness and criteria for grading motion sickness severity (Graybiel et al. 1968)

Category	*Pathognomonic (16 points)*	*Major (8 points)*	*Minor (4 points)*	*Minimal (2 points)*	*AQS (1 point)*
Nausea syndrome	Vomiting or retching	Nausea II, III	Nausea I	Epigastric discomfort	Epigastric awareness
Skin		Pallor III	Pallor II	Pallor I	Flushing
Cold sweating		III	II	I	
Increased salivation		III	II	I	
Drowsiness		III	II	I	
Pain					Headache
Central nervous system					Dizziness Eyes closed > II Eyes open III

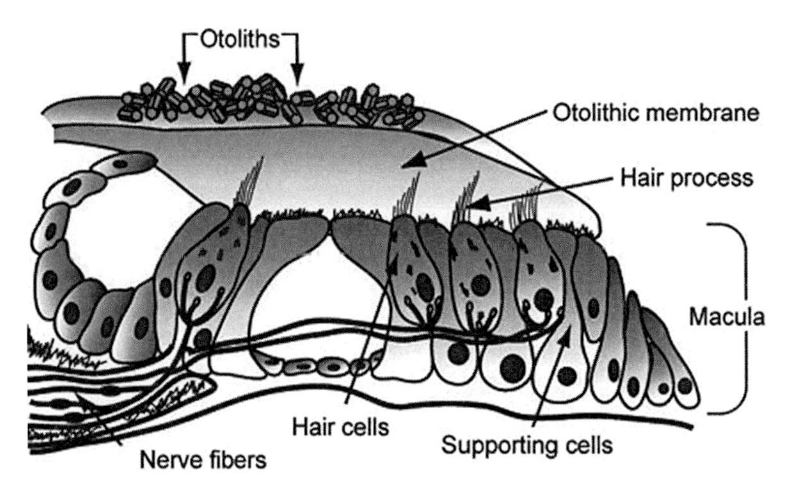

7.22 Vestibular system. Credit: NASA

pre-flight adaptation training techniques such as virtual reality. The theory behind this type of training (see sidebar) is that devices such as virtual reality can simulate the sensory realignment that occur in microgravity that causes SMS: by repeatedly exposing astronauts to unusual environments, it should be possible for crewmembers to encode and adapt to these challenging stimuli.

Pre-flight Virtual Reality Training

Astronauts engaged in this type of training typically perform standard mission tasks, such as navigation or switch activation, in multiple orientations in a virtual environment in the hope that they will retain the skills acquired when it comes to the real thing. Over the years, the training has proven to be quite effective, with the number of crewmembers suffering from SMS symptoms reduced by half. Another similar training method is autogenic feedback training (AFT), which employs psycho physiological countermeasures to condition crewmembers to voluntarily control their physiological responses. The training, which takes place over several days, involves a lot of repetition and practice but, at the end of the conditioning, most people are able to exert a greater control over the physiological responses to motion stimulation.

Pharmacotherapy

If virtual reality and AFT fail, what can you do? Well, there is always the pharmacotherapy option. To reduce the effects of SMS, astronauts can be prescribed antihistaminic agents such as Meclizine, anticholinergic agents such as Scopolamine, and antihistaminic agents with anticholinergic effects such as Promethazine and Diphenhydramine. Unfortunately, these drugs cause side effects such as drowsiness and lack of concentration. Imagine spending US$150,000 on a trip of a lifetime and falling asleep! That would be a big downer! But being sick and spending the entire flight looking into the contents of a rapidly expanding vomit bag isn't much fun either, so you need to do something that guarantees you an emesis-free flight and drugs might just be the answer. The trick to offsetting the side effects we just mentioned is to prescribe more drugs, only these concoctions counter the side effects of the first. For example, to counter the drowsiness induced by Scopolamine, Dexamphetamine is taken and, to offset the sluggish behavior caused by Promethazine, Ephedrine is given. While these combinations work for most, taking drug cocktails isn't for everyone: some people taking these drug cocktails have reported feeling very jittery and others have experienced symptoms of rapid heart rate. But these drugs can be more effective if the intranasal mode of administration is taken. For example, it is known that the intranasal administration of Chlorphedra has no negative effects on cognitive performance. This is because the nasal method offers a more direct route to the central nervous system and bypasses the metabolism in the gut wall. The only noticeable side effect is nasal irritation.

Perception

In addition to the problems of barfing, suborbital astronauts must also contend with the effects on their perception, which is disrupted as a result of the illusory changes during the flight. For example, there may be many who will find the inversion illusion troubling, which means it will be important to focus on strong orientation cues and secure restraints, the latter of which won't be a problem in the Lynx because the passenger and pilot are strapped in. But, even though our suborbital scientist will be strapped in, disruption of perception will cause a reduction in productivity, impaired performance, and an increased risk of mishap. So what can be done to deal with all these disruptions? Well, there are some strategies which are outlined below.

Strategies to minimize sensorimotor disruptions:

1. Adaptation and pre-adaptation

   The key to adaptation is repetition, which is why suborbital Lynx pilots will have an advantage over their passengers because they may be flying several times a week. This process of adaptation and re-adaptation, combined with suitable cognitive training, will benefit the pilots to the extent that they probably won't have to rely on pre-adaptation and/or pharmaceuticals, which is a good thing given some of those side effects discussed earlier. For suborbital scientists who may be flying just once, the best way of adapting is by using parabolic flight (see later in this chapter). How many? It's difficult to say because there is quite a range of rates of adaptation, but generally it takes up to three days or three to four minutes at zero-G per day for sensorimotor changes to develop. This means you will need to fly one parabolic flight sequence per day for three days before your Lynx flight. That's an expensive way to adapt! An alternative is using short-radius centrifugation and/or unusual attitude training in a high-performance jet fighter, but there is still the cost issue.

2. Re-adaptation

   Now you may be wondering how long the sensorimotor changes (see sidebar) that occur during the adaptation phase last. After all, there is no point embarking on a parabolic flight three weeks before your Lynx flight if those adaptive changes are only preserved for two weeks. Well, scientists have studied this and found that re-adaptation occurs within the first five days but then drops off after that time frame. So, if you want to ensure maximum re-adaptation, schedule your parabolic flight within five days of your Lynx flight. And, if you're lucky enough to be flying a second flight, it will make sense to schedule it as soon after the first as possible, because those sensorimotor adaptations begin to degrade after a few weeks.

Discombobulation

As the Lynx racks up flights, flight surgeons will get a better idea of the sort of sensorimotor disruptions that will affect suborbital astronauts. This database will be generated by the results of pre-screening and pre-, in-, and post-flight neurological function tests and assessment of any neurovestibular problems experienced during missions. Some of these conditions may have no bearing on flight safety, while others, such as benign paradoxical positional vertigo (BPPV) and vestibular migraine, may have longer-lasting consequences. No doubt, this database will form the basis of a research endeavor to correlate sensorimotor function with astronaut performance, the outcome of which will be a series of countermeasures specific to the deficits observed.

RADIATION

Astronauts working on board the ISS are constantly bombarded by radiation. In fact, the radiation environment in low Earth orbit (LEO) is so dangerous that space agencies have had to impose career limits. First there is the cosmic electromagnetic radiation, which includes gamma ray bursts, and then there is solar particulate radiation that includes solar flares. Being exposed to these types of radiation can cause long-term damage which is why the ISS is fitted with shielding but, for flights on board the Lynx, radiation shouldn't be an issue because these missions will only reach comparatively low altitudes (Table 7.4) and exposure will be measured in minutes and not hours or days. Still, it's helpful to understand the radiation environment, if for no other reason than to put your mind at rest when you strap into the right seat. As a radiation reference point, the general population is exposed to a background dose of radiation of about two to three millisieverts (mSv) per year. To put that in perspective, that annual radiation dose equates to the radiation exposure corresponding to around 300 suborbital flights per year. Another equivalent comparison is to use the X-ray, which corresponds to around 11 suborbital flights. So, even if you're a hot-shot pilot who flies into space almost every day, radiation is an insignificant issue. For those who like to put a figure on risks, I've included a table (Table 7.5) showing the dose rates that are assumed for a flight on board the Lynx

As you can see, you have nothing to worry about, even if you're rich enough to be flying every week. In fact, you would have to fly 188 flights per year before exceeding the dose limits for the general public. Now, if you happen to be a NASA astronaut, those limits

Table 7.4 Terrestrial dose rates for radiation (equivalent whole-body dose)

Altitude (m)	*Equator (mSv/day)*	*55° latitude (mSv/day)*
5,000	0.5	0.8
10,000	2	4
15,000	4	12
20,000	4	4.5
25,000	4	15
30,000	3	14
40,000		12.5
50,000		12

Table 7.5 Equivalent whole-body dose rates for radiation during suborbital flight

Altitude (m)	*Equivalent whole-body dose (mSv/day)*
7,500	0.17
15,000	0.24
30,000	0.29
45,000	0.32
60,000	0.35
90,000	0.40
120,000	0.46

are even higher because this group's annual dose limit is 0.5 Sv. That equates to 94,000 suborbital flights per year. Like I said, radiation is not something you have to worry about if you are flying on the Lynx! Let's move on.

PARABOLIC FLIGHT

After all this talk of G-LOC (sidebar), and motion sickness, you may be wondering whether spaceflight is your cup of tea, so why not take a test run of sorts? We're talking about parabolic flight (Figure 7.23). Although a zero-G flight only provides 20–24-second snapshots of what being weightless feels like, it is an invaluable training tool and also a great way of knowing whether your body is up to the real thing; if you spend most of your time projectile vomiting, then perhaps suborbital flight isn't in your wheelhouse. But, if you have a blast, then you can look forward to the real thing.

7.23 Parabolic flight. Credit: ESA

G-LOC and Geasles

Tissue ischemia is a term used by physiologists to describe insufficient blood flow and it is a familiar term in the realm of acceleration physiology because it is ischemia that is the most significant effect of G. Since the eye's retina is so sensitive to hypoxia, symptoms of sustained and/or increased G are usually manifested visually. As the Gs pile on, retinal blood pressure falls below the eye's globe pressure and blood flow to the light-sensing receptors in the retina also falls, with the result that vision is lost progressively from the periphery. Tunnel vision progresses to grayout and then to blackout – a condition that is termed full retinal ischemia. The end result is G-LOC, a condition that may be accompanied by myoclonic convulsions, amnesia, and general discombobulation.

It isn't just the circulatory system that suffers during sustained and increased G because the respiratory system also takes a hit. During +Gz, respiration is disrupted as the increased pressure of G collapses the small air sacs in your lungs, which obviously makes it difficult to breathe. And then there is the condition known as G-measles, or Geasles, which is caused by ruptured capillaries, which in turn results in unsightly red blotches.

Before we discuss the benefits of parabolic flight training, it's worthwhile reminding ourselves of the distinction between free fall and weightlessness. Way back when the Shuttle was flying at an altitude of around 300 kilometers, gravity as measured on board the orbiter was only slightly less than measured at sea level (9.37 meters/second^2 on board the Shuttle versus 9.81 meters/second^2 at sea level). So, when terms such as *microgravity*, *zero-G*, and *weightlessness* are used to describe gravity in orbital flight, these terms are technically inaccurate because spacecraft are constantly falling towards Earth under the force of gravity; the reason vehicles remain in orbit is thanks to their velocity, and the reason astronauts perceive themselves as being weightless is due to the fact that they are falling under the influence of the gravitational field of the spacecraft. In reality, astronauts are in a perpetual state of free fall and this is something that can be replicated closer to Earth, albeit for much shorter periods. In parabolic flight (Figure 7.24), an aircraft flies a trajectory that provides up to 25 seconds of free fall.

Parabolic flight as a training tool for astronauts has a history dating all the way back to 1950 when the technique was tested by ace test pilots Chuck Yeager and Scott Crossfield at Edwards Air Force Base. Over the years, the technique was refined with the arrival of the F-94 fighter that permitted up to 30 seconds of free fall, and government organizations began to adopt parabolic flight programs for astronaut training and research. Here's how NASA's C-9B aircraft flies its parabolic trajectories. Once it reaches 350 knots of indicated airspeed (Mach 0.83) and an altitude of 7,300 meters, a gradual climb is initiated at full thrust, thereby generating vertical speed without sacrificing airspeed. During the gradual climb, the G-meter reads 1.5 Gs, but this reading increases to 1.8 Gs as the pitch angle increases to 45° (the "pull-up"). At 225 knots of indicated airspeed, with the aircraft closing in on an altitude of 10,000 meters, the pilots begin the zero-G parabola by pushing

7.24 European Space Agency (ESA) astronauts training in parabolic flight. Credit: ESA

forward on the control yoke. This lowers the angle of attack of the wings, which in turn reduces wing lift. As the power is simultaneously reduced, airspeed falls as the aircraft reaches the top of the parabola, which it reaches at 10,000 meters (Mach 0.43 or 140 IAS/245 TAS – this speed isn't much faster than the stall speed). This is the fun part when passengers and wannabe astronauts start floating around the cabin. It is also the point at which newcomers to the world of zero-G realize that working in weightlessness can be a bit of challenge because faulty proprioception leads to target overshoots, limb control disruption, and slower limb movements. Sadly, the parabola only lasts about 24 seconds, after which the aircraft pitches down (the "push-over") and the Gs ratchet up to 2 Gs until the aircraft levels off for the longitudinal component before starting the next parabola.

8

STEM

Credit: Steve Heck

"One of the most exciting developments in commercial spaceflight is that it will soon be possible for students across the country to send their ideas and experiments to suborbital space. Steve Heck and his colleagues at The Arête STEM Project have pioneered student suborbital payloads, and the program they started in Cincinnati, Ohio is the gold standard."

Michael Lopez-Alegria, President of the Commercial Spaceflight Federation

E. Seedhouse, *XCOR, Developing the Next Generation Spaceplane*,
Springer Praxis Books, DOI 10.1007/978-3-319-26112-6_8

STEVE HECK

Meet Steve Heck. These days, Steve is an educator astronaut – a new job title created thanks to the increasing emphasis on STEM (science, technology, engineering, and mathematics) in the world of commercial suborbital spaceflight, but before he began encouraging high-school students to fly their experiments in space, Steve was a US Air Force (USAF) pilot. During his 20+ years of service Steve rose to the rank of Lt Colonel while amassing more than 2,700 flight hours on all sorts of aircraft. Serving as a Command and Instructor Pilot, Steve combined his love of flying with his enthusiasm for education – work that culminated in his being nominated for one of President Bush's "Points of Light" awards. In addition to receiving five Meritorious Service Medals while in the USAF, Steve was also an Outstanding Graduate from the USAF's Air War College. After retiring from the military, Steve went hard to work as an educator, working closely with NASA as an Astronaut Educator in the Citizens in Space Program – work that garnered him a NASA Endeavour Fellowship. If you're looking for one of the catalysts responsible for putting STEM on the suborbital radar, then look no further than Steve. In 2013, Steve graduated from NASTAR's Suborbital Scientist Training Program (SSTP) and then went about creating the Arête STEM Project, the first program of its kind that aligns the commercial spaceflight industry with K-12 Education. It's proven to be a very, *very* popular match. During 2013, 2014, and into 2015, Steve presented the concept at several schools in Ohio, pitching the opportunity for school children to fly their science experiments in space. For free! How is this possible?

In 2012, the State of Ohio inducted Steve into the Ohio's Veterans Hall of Fame for his exemplary service to the US and excellence in education. In 2013, Steve earned his civilian astronaut wings after graduating from NASTAR's SSTP. Steve works with the Suborbital Applications Research Group (SARG) as its national K-12 Education and Public Outreach Representative, and his Arête STEM Project brings together the Commercial Spaceflight Industry and K-12 Education: the goal of this project is to help students design STEM experiments to fly into space while making spaceflight available, at no cost, to schools.

STEM

For those of you unfamiliar with STEM, here's some background. Based on the four disciplines of science, technology, engineering, and mathematics, the STEM curriculum aims to educate students using an interdisciplinary and applied approach. So, instead of teaching the disciplines separately, the STEM world is one of integration and cohesion. In the US, STEM education gained traction during the Obama Administration as a result of fewer students pursuing STEM subjects after graduating from high school. To motivate students to excel in STEM subjects, the Obama Administration kick-started the 2009 "Educate to Innovate" program and proceeded to invest federal funds in STEM education together with STEM grant selection programs and research programs that supported STEM education. The programs were so successful that the Obama Administration's 2014 budget allocated US$3.1 billion in federal STEM programs. The innovative aspect of STEM is the blended learning approach that demonstrates to students how the scientific method can be applied to all manner of real-world applications. And this type of learning begins at elementary school, where teachers pique students' interest by making them aware of STEM fields and occupations. By the time students arrive in middle school, their course material becomes more challenging and, by the time they enter high school, students are familiar with the courses and pathways available in STEM fields and the various occupations that require a STEM background.

NASA and STEM

Given the lack of interest in STEM subjects, it isn't surprising that NASA has been very active in promoting the programs across colleges and technical schools. After all, it's almost impossible to get a job with the agency without at least one of the STEM subjects. In 2014, NASA's Office of Education awarded US$17.3 million via the National Space Grant and Fellowship Program to increase awareness of STEM subjects. For example, the California Space Grant Consortium has a multi-faceted program that develops STEM courses for students with a special focus on such space-related areas as small satellites and near-space ballooning. A similar approach is taken by the Colorado Space Grant Consortium that encourages students to apply their STEM knowledge to activities such as building and launching high-altitude balloons.

Arête STEM

As educational initiatives go, Steve's Arête STEM Project (*www.arete-stem-project.org*) is about as cutting-edge, unique, and bold as they come. Designed to help students develop the skills necessary to compete for jobs in an increasingly complex workplace, Arête STEM focuses on the STEM design process with a unique twist: it aims to excite and challenge students by offering school kids (Figure 8.1) the opportunity to design, build, and fly their science in space. Key to the educational program is Arête STEM's relationship with XCOR and their Lynx. Here's how it works.

If you happen to be a student in Cincinnati or the surrounding area, your school can start your space adventure by reserving a payload slot on the Lynx. Most likely, this will be a slot for the flight of a payload that will be housed in the Cube Carrier and the payment (around US$3,000) will be paid for by a community partner who can offset the cost as a tax write-off. With their reservation secured, students can then go about the business of readying their payload for flight via the Engineer Design Process. Along the way, the students are given assistance and their progress monitored to ensure milestones are reached and goals met. Then, 90 days before launch, students are requested to provide a progress check before completing the final phases of test and analysis before their payload finally makes its way into the Cube Carrier. The next milestone is 60 days before launch, when students conduct simulated flight testing of their experiment, after which data are reviewed and any redesigns are taken care of. Then, 30 days before launch, all safety checks are conducted and the payload is shipped to Midland for integration into the Cube Carrier on board the Lynx. The flight takes place, data are returned to the students, data are analyzed,

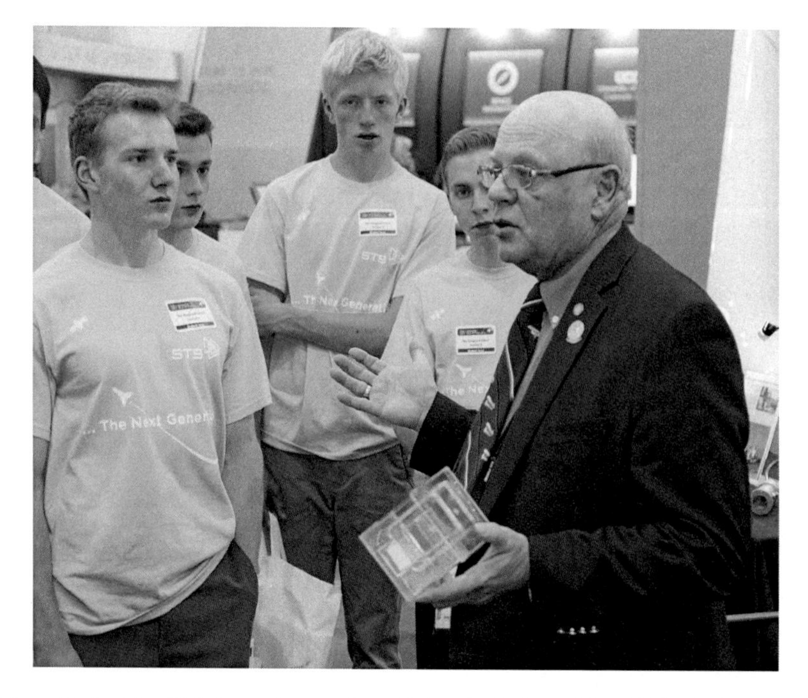

8.1 Steve Heck at the 34th Space Symposium, Colorado Springs, May 2015. Credit: Steve Heck

and a research paper is written, submitted, and published. That's pretty cool if you happen to be a high-school student. After all, how many kids will be able to list "science experiment flown in space" on their resume or their university application? But Arête STEM isn't just about students (Figure 8.2) flying their science projects in space; it is also about creating a network of mentors who can work with students on their projects and fostering a business, community, and school relationship to enable these missions. To date, Steve's initiative has been very successful. Take Milford School, for example.

> "My favourite part was the adrenaline of waiting to see who was going to get to go into space."
>
> *Fifth-grader Ty Dominguez of Meadowview Elementary whose experiment will investigate how microgravity affects an egg that hatches on Earth*

Milford School's "Right Stuff"

Before explaining how Steve's initiative made its way into the mainstream education curriculum, it's worth noting the geographical significance of the Arête STEM's project location, since Ohio was home to not only the Wright brothers, but also astronauts John Glenn and Neil Armstrong. That historical footnote probably helped the project gain a little traction, but perhaps one of the biggest supporters who helped get Steve's project kick-started

8.2 Steve Heck with students in front of the Lynx mock-up at the 34th Space Symposium, Colorado Springs, May 2015. Credit: Steve Heck

was Superintendent Robert Farrell, who provided many of the tools necessary to make the enterprise work. Another important factor was the Duke Energy Foundation, a community partner of the Milford Schools Foundation that raised money for the STEM education program. With the funding and support in place, it was time to train the teachers, after which the Milford School district's fifth-grade students visited iSPACE, a non-profit dedicated to creating STEM awareness. With the students up to speed on STEM, 137 teams of fifth-graders from the district's six schools were formed to design an experiment. The mission was simple: each team was given a 10 × 10 × 10 centimeter cube (Figure 8.3) and was told they could fly pretty much anything they wanted, as long as it fitted into the cube.

> "I remember where I was when John Glenn orbited the earth in his Mercury capsule. I remember sitting in front of a black and white TV watching Neil Armstrong walk on the moon. We're going to recreate that time when kids were excited about science."
>
> *Steve Heck being interviewed about STEM education at Milford School*

Other than the confines of the cube, the 580 students who comprised the 137 teams were given few constraints to design and test their experiments. Some students made use of littleBits kits to develop their project and some, who needed a power source for their experiment, went ahead and tested which power connections would best fit into the cube. After much comparative testing, it was found that the p2 Coin Battery worked best because it was compact and rechargeable. For those who needed various modules held in place, mounting boards were cut to size using a Dremel. In short, ingenuity was the name of the

8.3 Arete's Cube Carrier. Credit: Steve Heck

game and competition was intense, but eventually teams were winnowed down to Best in Class entries by the teachers. The top 10 teams were:

- "Egg in Microgravity": Team 13 at McCormick Elementary (Aaron Coors, Ethan Creer, Seth Eastham, Collin Murphy, and Pierce Will)
- "Neil Armfish": Team 6 at Pattison Elementary (Ethan Holman, Vishnu Rajkumar, A.J. Evans, and Dylan Mullarkey)
- "Egg Project": Team 21 at Meadowview Elementary (Nick Bohlander, Ty Dominguez, Heidi Cook, Janie Tudor, Michael Cotton, and Gabe Carman)
- "Gravitational Water Transfer": Team 8 at Boyd E. Smith Elementary (Mia Dearing, Anya Moeller, Josh Panko, and Joel Sagraves)
- "Ladybugs in Microgravity": Team 15 at Mulberry Elementary (Natalie Earl, Chase Lemle, Meghan Gentry, Ryland McGahey, and Andrew Palmer)
- "Compass Magnetic Field in Salt Water, Oil and Air": Team 2 at Seipelt Elementary (Liddy Dow, Tiffany Lau, Spenser Hore, and Ian Golden)
- "Microgravity Density Experiment": Team 11 at McCormick Elementary (Malachi List, Kyle Dolby, Johnny Mei, Brett Rininger, and Harley Healey)
- "Testing the Heart Rate in Space": Team 18 at Meadowview Elementary (Samantha Jones, Gabriel Ditullio, Jocelyn Howard, and Nicholas Luciano)
- "Substances Mixture": Team 7 at Pattison Elementary (Riley Eggemeyer, Laura Winterod, Olivia Ossola, and Carley Eggemeyer)
- "Worms in Space": Team 14 at Mulberry Elementary (Hector Camacho, Jacob Bateman, Jaquey Bean, and Emily Nelson)

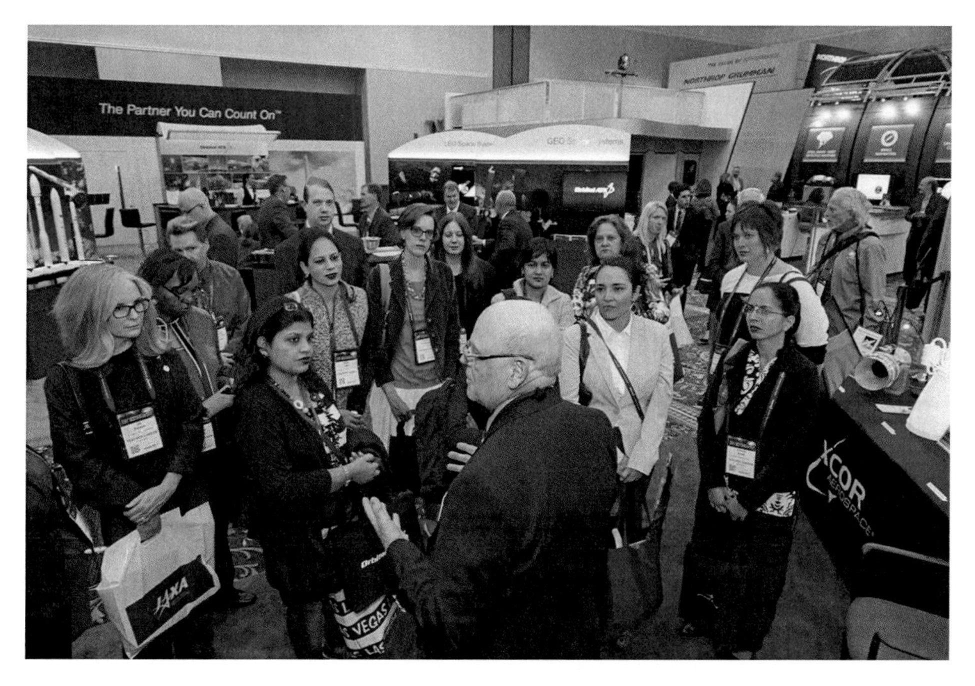

8.4 Steve Heck and students. Credit: Steve Heck

At the time of writing, the students are waiting for their slot on the flight manifest and Steve's initiative has been adopted by other schools not just in Ohio, but in several other states too. The potential of this space-themed STEM initiative is huge, not only because it has changed how students and teachers view STEM education (Figure 8.4), but also because it helps students develop so many new skill sets (see sidebar).

How Space-Themed STEM Projects Help Students

- Students become familiar with how to revise experimental designs based on peer review
- Students analyze the positive and negative effects of technology on the environment
- They revise designs used to resolve a problem based on peer review
- They select the tools necessary to collect data when conducting investigations
- They assess observations and measurements made by other students and identify reasons for errors
- They identify hazards associated with an investigation
- They adopt an evidence and observation-based approach to communicate the results of investigations
- They can explain how conclusions changes as new knowledge is gained
- They develop evidence-based explanation to defend outcomes

CITIZENS IN SPACE

Citizens in Space (*www.citizensinspace.org/*) is a non-profit endeavor designed by the US Rocket Academy to enable scientists to fly in space. As part of the project's first phase, the group has a contract for 10 suborbital spaceflights with XCOR. Experiments, which will be housed in the Lynx's Cube Payload Carrier, will be flown free of charge provided the experiment is licensed as open-source hardware. Citizens in Space is also in the business of selecting and training up to 10 citizen astronauts, who will fly as payload operators. To date, the following experiments (see sidebar) have been announced:

- Angelicvm Aerospace Foundation of Santiago, Chile: "Crystallization Rates in Microgravity"
- Bishop Planetarium at the South Florida Museum in Bradenton, Florida: "Microgravity Water Electrolysis Optimization"
- CD-SEAS of Honolulu, Hawaii: "Effectiveness of Anti-Microbial Coatings in Microgravity Conditions"
- Florida International University of Miami, Florida: "Regolith Compression Mechanics in Reduced- and Micro-Gravity"
- Flightsafety Makers of Columbus, Ohio: "Characterization of Local Inertial Loading and Comparison with Avionics Data"
- Syncleus of Philadelphia, Pennsylvania: "Realtime Payload Conditions Monitoring"
- NewSpace Farm LLC of Seattle, Washington: "Microgravity Botany Pod Hardware Evaluation"
- The Pinkowski Group of Montrose, Pennsylvania: "Concentration Gradient Equalization Rates"
- Terran Sciences Group of Orlando, Florida: "Inter-Payload Heat Transfer Characterization"
- Texas Southern University of Houston, Texas: "Non-Fick Diffusion in Microgravity"
- Students for the Exploration and Development of Space at the University of Central Florida in Orlando: "Hydrophobic Coating Effectiveness for Space Applications"
- University High School of Orlando, Florida: "Investigation of Regolith Hydration in Zero Gravity"

[sidebar] If you are interested in submitting an experiment, then just contact Dr. Justin Karl by emailing experiments@rocketacademy.org. Good luck!

It's a great initiative because not only does the Citizens in Space (see sidebar) program develop future scientists, it is also the catalyst for launching research programs for schools, and also creates a valuable network of mentors who work with students on their space projects. And the really cool aspect of the program is that fundraising efforts generally offset the cost of the missions. Oh, and the cost. You may be thinking that flying a science experiment will be awfully expensive, but it really depends on the experiment. Let's begin with a reference point. To fly a compact CubeSat, you might need to cough up between US$5,000 and US$7,500 to actually build the payload and another US$40,000 or so to launch it. That's too much for school students. But, with the Citizens in Space program, kids can fly their stuff in the Lynx Cube Payload Carrier, which slots in snugly behind the pilot's seat. And, as long as you can stuff your experiment or payload into the 10 × 10 × 10 centimeter cube, the cost is around US$3,000.

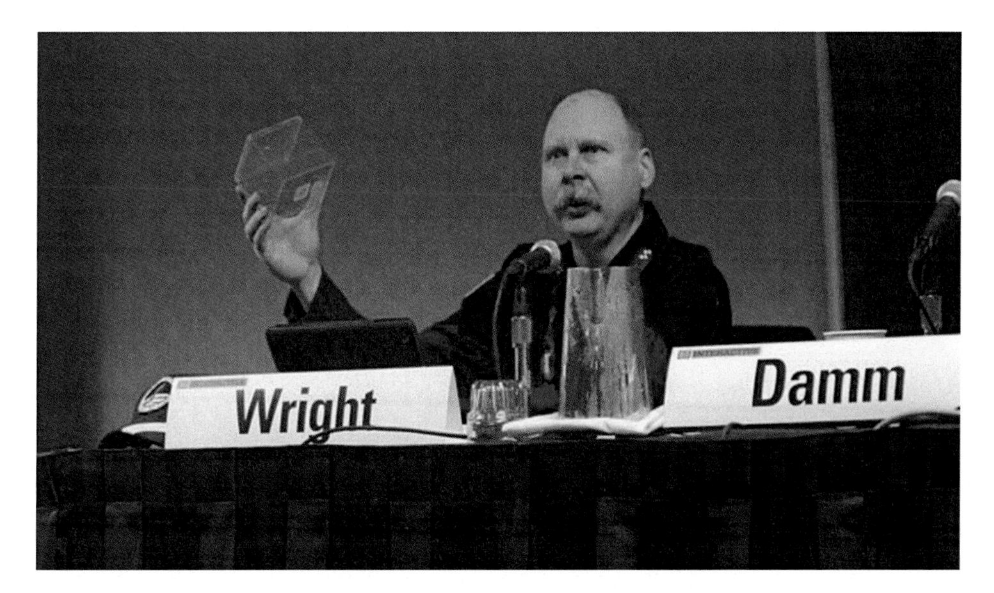

8.5 Ed Wright. Credit: Henrik Brattli Vold/NRK

Citizens in Space

Chairman of the US Rocket Academy is Edward Wright (Figure 8.5), who is also the project manager for Citizens in Space. In an earlier life, Ed spent 20 years working in the computer software industry and as president of X-Rocket, LLC – a company that operated the MiG-21 fighter jet. Nowadays, he spends his time leading Citizens in Space, recruiting, selecting, and training citizen astronauts, and promoting the numerous science experiments destined to fly on suborbital rockets.

The commercial suborbital industry offers some absolutely game-changing opportunities for NASA's research and its education missions. These new flight systems, which offer a powerful combination of high-frequency spaceflight at low cost, are going to open a lot of exciting doors for technology testing, for Earth science, for microgravity sciences, for life sciences, experiment technology readiness level-raising, training, education, public outreach, and other areas, too.

Alan Stern, Space News *interview by Brian Berger, 25 July 2011*

Suborbital Applications Researchers Group

Around the same time as Citizens in Space was taking shape, a similar initiative was taking place under the direction of the Commercial Spaceflight Federation (CSF) which created the SARG in 2009. Comprising scientists and educators interested in realizing the research and education potential of the new fleet of suborbital vehicles, SARG (see sidebar) has been instrumental in promoting the potential of Research and Education

Missions (REM) on board these vehicles. To that end, the group created the SARG Ambassadors Program, which is tasked with educating the suborbital operators about the potential of REM missions, encouraging NASA and other agencies to fund REM missions and to increase awareness of commercial vehicles in the REM communities.

SARG Members

Dr. Steven Collicott, SARG Chair, Purdue University – Microgravity Physics
Dr. Makenzie Lystrup, Vice Chair, Ball Aerospace – Planetary Science
Dr. Sean Casey, Silicon Valley Space Center – Astrophysics
Dr. Joshua Colwell, University of Central Florida – Microgravity Physics
Dr. Daniel Durda, Southwest Research Institute – Planetary Science
Mr. Steve Heck (USAF-ret), Citizens in Space – Education and Public Outreach
Dr. Anna-Lisa Paul, University of Florida – Life Sciences
Dr. Mark Shelhammer, Johns Hopkins University – Space Life Sciences
Dr. H. Todd Smith, JHU Applied Physics Laboratory – Aeronomy
Dr. Stan Solomon, National Center for Atmospheric Research – Atmospheric Sciences
Mr. Charlie Walker – Human Spaceflight

You may be wondering why someone as multi-talented as Stern never became an astronaut. The reality is that he made the cut on every level – except the medical (due to a detached retina). Since he was denied working in low Earth orbit (LEO) as a NASA employee, Stern applied his considerable will and tenacity as a project scientist, working as principal investigator on a number of programs within NASA (see sidebar) before staking out a new position at the Southwest Research Institute (SwRI) in San Antonio, Texas, and then Boulder, Colorado, in 1994. Back in those days, SwRI's Boulder outpost comprised just Stern, a researcher and a secretary. Fast forward 20 years and the Boulder operation employs more than 50 scientists and takes in more than US$40 million a year.

Alan Stern and NASA's New Horizons Mission

Alan Stern (Figure 8.6) is one of leading advocates of suborbital research, but you've probably heard of him because of his role as Principal Investigator on NASA's New Horizons mission. The US$700 million mission is one that Stern worked on for the best part of 25 years and, for those of you who followed the mission in the summer of 2015, you will remember that it was an unmanned mission unlike any other: one that had the eyes of the world following every trajectory correction and maneuver until the spacecraft's successful encounter and fly-by of our solar system's largest dwarf planet.

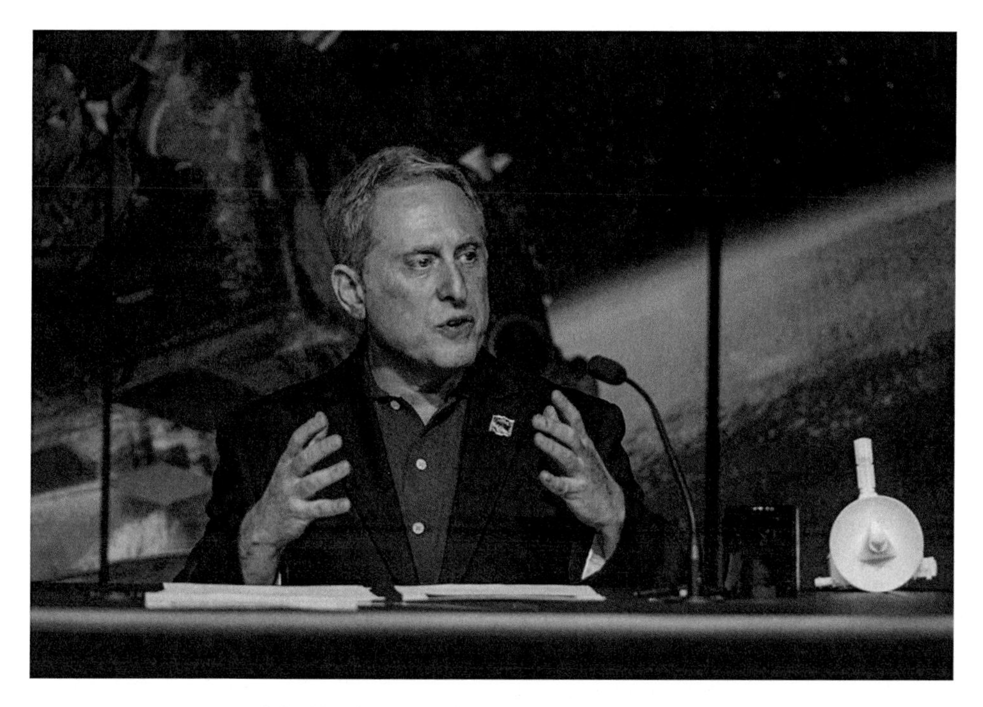

8.6 Alan Stern. Credit: NASA/Aubrey Gemignani

Stern is not only a guy who got things done in the world of NASA, but also someone who has been a catalyst for so much in the suborbital spaceflight industry. It was Stern who, in February 2011, brokered the first deal of its kind, when SwRI bought six tickets on XCOR's Lynx Mark I vehicle. Since then, Stern and his colleagues, Dan Durda and Cathy Olkin, have completed suborbital flight training at NASTAR and on board the F104. In between brokering suborbital flights (SwRI also purchased flights on board Virgin Galactic's SpaceShipTwo) and training as a commercial astronaut, Stern has somehow managed to find time to kick-start the Next Generation Suborbital Researcher's conference and consult for commercial space companies such as Blue Origin.

9

How to Fly

Credit: XCOR

E. Seedhouse, *XCOR, Developing the Next Generation Spaceplane*,
Springer Praxis Books, DOI 10.1007/978-3-319-26112-6_9

> *"Being the first Irishman in space is not only a fantastic honor but pretty mind-blowing. The first rock astronaut space rat! Elvis may have left the building but Bob Geldof will have left the planet! Wild! Who would have thought it possible in my lifetime."*
>
> Bob Geldof, September 2013

Yes, it won't only be Scientist-Astronaut Project Polar Suborbital Science in the Upper Mesosphere (PoSSUM) types who will be flying on the Lynx – there will be rock stars, porn stars, DJs, and Victoria's Secret models, too. But, before we profile some of the more famous names slated to take the right seat of the Lynx, let's take a look at the three astronaut programs XCOR has in place for those with US$100,000 to spare.

First there was the Founder Astronaut Program. I say that in the past tense because this program is sold out! This program was an exclusive program that was offered to the first 100 passengers, which included Victoria's Secret model Doutzen Kroes (Figure 9.1) and DJ Armin van Buuren (Figure 9.2). So, if you wanted to be among the first 100 to fly on the Lynx, too bad! But you can still sign up for the Pioneer Program.

9.1 Doutzen Kroes at the Cannes Film Festival. Credit: Georges Biard

9.2 Armin van Buuren at TomorrowWorld, September 2013. Credit: MIXTRIBE

To sign up for the Pioneer Program, simply visit the XCOR website and be prepared to cough up a 50% deposit – that's US$75,000. The rest must be paid within three months of your flight. The third program offered by XCOR is the Future Astronaut Program. After all the Founder Astronauts have flown, it will be the turn of this group of astronauts. The price is still US$150,000 though. Don't have that kind of money? Not to worry, because there are other ways of getting your ride to space without paying a dime. But, before we look at what those options might be, let's profile some of XCOR's more recognizable future spacefarers.

XCOR'S RICH AND FAMOUS

"Perspective is one thing, the beauty is the other, the thrill of it another and the f**k off adventure of it."

Bob Geldof, after being asked why he decided to take a trip to space

9.3 Bob Geldof at the 2014 One Young World Conference in Dublin. Credit: Stefan Schäfer

Many years ago, when I was at school, Bob Geldof (Figure 9.3) was the front man for the Boomtown Rats, a phenomenally successful Irish new-wave band that had a series of iconic hits, including "I Don't Like Mondays," "Rat Trap," and "Banana Republic." Fast forward 30 years and Bob Geldof is now better known as a philanthropist with an honorary knighthood and a pair of Nobel Peace Prize nominations under his belt. Very soon, he will be able to add "Astronaut" to his resume. In all likelihood, Ireland's icon of social responsibility will also probably be the first rock musician *and* first Irish citizen in space (there are two Irish citizens – Bill Cullen and Tom Higgins – booked on Virgin Galactic's SpaceShipTwo but, after SS2's accident in October 2014, it is likely that XCOR will be first into space).

> "When I was a kid, I saw the movie 'SpaceCamp,' and it was such an awesome film. And I didn't even know that that was a real place that people could actually go and train to be an astronaut. To say, 'Yeah, I'm an astronaut' and really get my wings, that's thrilling."
>
> *Porn star CoCo Brown, aka Honey Love*

In addition to carrying the first Irish citizen and first rock star into space, XCOR will have the distinction of flying another first when the Lynx takes off with CoCo Brown in the right seat. CoCo Brown is a retired porn star who now works as a DJ. She had never

9.4 CoCo Brown at the 2013 AVN Awards in Las Vegas. Credit: Michael Dorausch, *michaeldorausch.com*

considered the possibility of becoming an astronaut until a concierge acquaintance slipped her an invite to a space luncheon. Intrigued, Ms. Brown (Figure 9.4) attended the luncheon and signed up.

> "I wonder if my boobs will float. Will they? There's no gravity. I never noticed it in the training. Do fake boobs do something weird in space? I don't know. Maybe I'll pop my boob out and take a photo of it with the Earth in the background."
>
> *CoCo Brown, former porn star and astronaut-in-waiting, pondering her in-flight to-do list*

> "Become an Axtronaut
> AXE Company Wants to Launch 22 People Into Space"

That was the headline that kicked off Unilever's AXE Apollo Space Academy in early 2013. The pharmaceutical behemoth that owns the men's personal care product company AXE had joined forces with Buzz Aldrin to promote an online contest that promised to select 22 winners to fly to space on board the Lynx.

> "Space travel for everyone is the next frontier in the human experience. I'm thrilled that AXE is giving the young people of today such an extraordinary opportunity to experience some of what I've encountered in space."
>
> *Buzz Aldrin, Apollo 11 astronaut and AXE spokesperson*

The contest was open to more than 60 countries and all that prospective space travelers needed to do to enter was to write a short essay about why they should be chosen to fly. It also helped if contestants could drum up a lot of online support because the winners were decided by online votes. After much online voting, a group of 109 AXE space cadets converged on AXE Space Camp in Florida at the end of 2013 to begin the final step of the year-long competition. There, they took part in a watered-down version of astronaut training – a program that included some zero-G time and a short flight on board a fighter jet. And, after some deliberation, 22 finalists were selected to fly on the Lynx. Those venturing into orbit received their tickets at a graduation ceremony at the Kennedy Space Center hosted by Buzz Aldrin.

> "The Apollo campaign was not only incredibly exciting for the AXE community; it energized and engaged an audience around the world. Through A.A.S.A. [AXE Apollo Space Academy], we're proud and excited to make history by giving guys and girls the ultimate chance to go to space – making space accessible in a way that hasn't been done before."
>
> *Matthew McCarthy, AXE's senior director of brand building*

The finalists were a mixed bunch that included 28-year-old Tim Gibson, an Australian who had earlier set his sights on becoming a pilot with the Royal Australian Air Force, and Tale Sundlisæter (Figures 9.5 and 9.6), aged 30, a senior engineer working for the Norwegian Air Force.

9.5 Tale Sundlisæter. Tale is a Ph.D. candidate at the Norwegian Defence University College. She serves on the Space Generation Advisory Council and has worked on various space projects, including support of camera development and testing for the ExoMars High Resolution Camera, the JUpiter ICy moon Explorer (JUICE), and the Hayabusa-2/MASCOT (Mobile Asteroid Surface Scout) MasCam. Credit: Tale Sundlisæter

9.6 Tale Sundlisæter. Credit: Tale Sundlisæter

FLY A PAYLOAD

If you don't have US$150,000 lying around to buy a ticket and you missed out on the competitions but would still like to participate in this impending suborbital era, how about launching a payload? This option will cost you as little as US$3,000. Or less if you happen to go after federal funding route because NASA has a program that funds these sorts of experiments. More of a key technology development pipeline link, the Flight Opportunities Program calls for payloads through an announcement of opportunity. In June 2013, the agency announced it had selected 21 space technology payloads for flights on commercial reusable launch vehicles, balloons, and a commercial parabolic aircraft. The selection represented the sixth cycle of the program, which has now facilitated more than 100 technologies with test flights – everything from systems that support cubesats to new sensor technology for planetary exploration. Of the 21 payloads selected in 2013, 14 will ride on parabolic aircraft flights, two will fly on suborbital reusable launch vehicles (sRLVs), three will ride on high-altitude balloons, one will fly on a parabolic flight and a suborbital launch vehicle, and another will fly on a sRLV and a high-altitude balloon platform. Although most of the payloads that have been selected to date are not suborbital, the program acknowledges the impending arrival of suborbital revenue flights and, once XCOR starts flying regularly, more payloads will be flown on sRLVs.

The main goal of the program is to develop and mature new technologies (Technology Readiness Level 4+). Selected proposals are offered a flight (or multiple flights) on a sRLV/parabolic aircraft, but are not provided with funding for payload development. The program funds more than a third of the proposals received and information about how to apply can be found at *https://flightopportunities.nasa.gov/*.

A similar program to the Flight Opportunities Program is NASA's Game Changing Development Program, which supports research from academia, industry, and governmental agencies. The program funds researchers to take their technology from a proof-of-concept stage (TRL 3+) to the component testing phase in an applicable environment. The program generally provides funding ranging from US$125,000 to US$500,000 for payload development in preparation for demonstration flights on sRLVs but does not guarantee flights for selected proposals; the next step for payload developers is to propose for a flight through NASA's Flight Opportunities Program.

If space and Earth sciences is your specialty, you might be interested in NASA's Research Opportunities in Space and Earth Sciences (ROSES) Program, which accepts proposals related to various Earth and space science initiatives within NASA. Proposals submitted to this program must be related to one or more of the following NASA Research programs: Heliophysics, Astrophysics, Planetary Science, and Earth Science. The ROSES program is open to groups including government agencies, private organizations, and nonprofits, and funding ranges from US$100,000 to US$1 million per year for a period of up to five years.

PROJECT POSSUM: A SUBORBITAL RESEARCH PROGRAM DESIGNED AROUND THE LYNX

> "This is an exciting time to be involved in spaceflight, and I feel very honoured and fortunate to be a part of such a special opportunity. This represents the next step in evolution and eventual routine access to space for all, and while fulfilling the dream of my lifetime, I will be able to help make a significant difference in our world through this groundbreaking research opportunity."
>
> *PoSSUM astronaut-candidate Pete Freeland*

Project PoSSUM (*www.projectpossum.org*) is an atmospheric science program that will be conducted on board the Lynx to study the upper atmosphere. The guy who devised this cutting-edge program and who also dreamt up the ingenious acronym is Dr. Jason Reimuller (Figure 9.7) (see sidebar), an ex-NASA system engineer and project manager

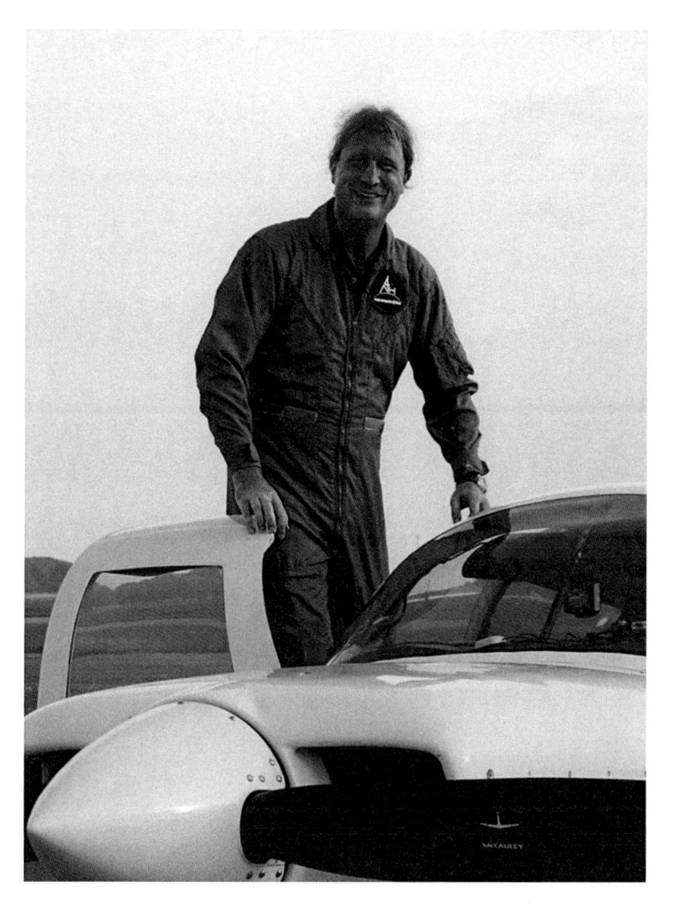

9.7 Dr. Jason Reimuller preparing to chase noctilucent clouds in a Mooney M20K aircraft. Credit: Jason Reimuller

based out of Boulder. Since being awarded a NASA flight opportunity award in 2012, Jason's innovative program has morphed into a fully fledged Scientist-Astronaut Program that is hosted at world-renowned Embry-Riddle Aeronautical University (ERAU) on Florida's Space Coast in Daytona Beach.

PoSSUM's first goal is to maintain a sustained research presence in the mesosphere – a region that has only been briefly transited in our forays to orbit. It is a region that harbors strange electrical phenomena and ionization that brings silence to vehicles re-entering through it. And, since it is an area too high to access by balloon or aircraft yet too low to access by orbital spacecraft, the PoSSUM research is perfect for the Lynx. During their flights, PoSSUM Scientist-Astronauts will focus their attention on noctilucent clouds, which are the highest clouds in Earth's atmosphere,

Project PoSSUM's Principal Investigator

Jason Reimuller is a research scientist with the Space Science Institute (SSI), President of Integrated Spaceflight Services (ISS), and is the Principal Investigator (PI) of Project PoSSUM, a research campaign that will study the polar mesosphere and noctilucent cloud structures. In addition to being one of the world's subject matter experts on the development of spacecraft egress training modules, training simulators, and analog space training capability, Jason also works as a commercial research pilot and flight-test engineer with GATS, Inc. The author of *Spacecraft Egress and Rescue Operations*, Jason served for six years as a system engineer and project manager for NASA's Constellation Program, leading studies on launch aborts, launch-commit criteria, landing conditions, post-landing and emergency crew egress trades, and propulsion options. He also led a NASA-funded flight research campaign to study noctilucent cloud time evolution, structure, and dynamics in Northern Canada as lead investigator and pilot-in-command, then further applied his background in airborne remote sensing to conduct research in glaciology to better understand the dynamic changes of the Greenlandic Ice Sheet as part of NASA's Operation ICE Bridge. Like so many over-achievers Jason holds a bunch of degrees that includes a Ph.D. in Aerospace Engineering Sciences from the University of Colorado in Boulder, an M.S. degree in Physics from San Francisco State University, an M.S. degree in Aviation Systems from the University of Tennessee, an M.S. Degree in Aerospace Engineering from the University of Colorado, *and* a B.S. degree in Aerospace Engineering from the Florida Institute of Technology. If you're interested in signing up for Project PoSSUM, just visit the website above or shoot Jason an e-mail at Jason.Reimuller@projectpossum.org.

forming at altitudes of 83 kilometers. These clouds are of keen interest in the study of global climate change and the Lynx is particularly well suited for PoSSUM research, since it can access the mesosphere by means previously unavailable. An additional goal of PoSSUM is to inspire the public and to communicate the science to broader audiences and the involvement of the trained human operators essential to conducting PoSSUM science missions (Appendix IV) provides a unique opportunity to engage the public through direct participation.

Why Study Noctilucent Clouds?

Noctilucent clouds (Figure 9.8) are widely studied by global climate scientists as well as developers of space vehicles designed to re-enter the atmosphere and planetary scientists studying atmospheres on other worlds. Climatologists are interested in the connection between noctilucent cloud presence and our changing atmosphere, seen by many scientists

9.8 Noctilucent clouds. Credit: NOAA

as probable indicators of long-term global climate change. When people are asked about climate change, the image that often comes to mind is of a polar bear stranded on a shrinking chunk of sea ice! But this image, and many others, is an effect of a warming climate: it is the atmosphere that is the cause of this change. And it is the atmosphere that we affect that causes all these changes, and the upper layers of the atmosphere are particularly sensitive to the man-made causes of climate change.

As the major observable phenomena in the mesosphere, noctilucent clouds are considered a sensitive indicator of global climate change, because a relationship has been observed between their presence and man-made industrial products: noctilucent clouds have been observed with increasing frequency over the last century, and this increase is seen by many scientists as indicative of long-term global climate change attributable to the increasing levels of atmospheric "greenhouse gasses." The theory is that, as carbon dioxide levels rise, the upper atmosphere cools and, as methane levels rise, more water vapor forms in the upper atmosphere. Therefore, the argument goes, the man-made causes of climatic change are believed to be directly related to the presence of noctilucent clouds. So, by better understanding noctilucent clouds, PoSSUM hopes to gain insight into the elements of global climate change believed to cause their expanding presence.

> "It is clear that noctilucent clouds are changing, a sign that a distant and rarefied part of our atmosphere is being altered, and we do not understand how, why or what it means .... These observations suggest a connection with global change in the lower atmosphere and could represent an early warning that our Earth's environment is being altered."
>
> *Dr. James Russell III, PI for NASA's Aeronomy of Ice in the Mesosphere (AIM) mission*

Another reason we are interested in noctilucent clouds is because they may pose a threat to re-entry vehicles. Since we don't know much about the mesosphere, NASA's manned space missions have enacted conservative flight rules regulating re-entry – constraints that may be irrelevant. But, through a better understanding of noctilucent clouds and the mesosphere, we may be able to design more operable space architecture.

Origins: The Noctilucent cloud Imagery and Tomography Experiment (NITE)

PoSSUM grew from an experiment conceived at the "Layered Phenomena in the Mesopause Region" (LPMR) conference held in Blacksburg, Virginia, in August 2011, where Jason was presenting the results of an airborne campaign that was the focus of his doctoral dissertation. Jason had designed and conducted an airborne flight research campaign in 2009, piloting a Mooney M20K aircraft over the Canadian subarctic imaging noctilucent clouds as NASA's Aeronomy of Ice in the Mesosphere (AIM) satellite passed overhead. The AIM satellite was the first satellite dedicated to the study of noctilucent clouds, and the synchronized imagery Jason obtained helped the AIM science team better assess the low-latitude (and most sensitive) cloud structures from a satellite with a primary mission to study noctilucent clouds from orbit. Following his presentation, Jason suggested to two colleagues, Dr. Gary Thomas and Dr. Dave Fritts (see sidebar), that they submit a proposal to NASA that would extend upon his dissertation experiment and use a spacecraft to study noctilucent clouds. Fortuitously, NASA's Flight Opportunities Program had just started accepting proposals to use reusable suborbital launch vehicles (rSLVs) to mature new space technologies. This was the perfect opportunity for Jason and his colleagues because they had their eyes on the Lynx which just happened to be custom-made for this type of study, since it was suborbital and affordable.

So Jason and his colleagues wrote a research proposal entitled the "Noctilucent cloud Imagery and Tomography Experiment (NITE)" and submitted it to NASA in December 2011. The premise was to use the Lynx's unique capabilities to build high-resolution imagery of noctilucent cloud layers as the vehicle transitioned through the layers. Accepted in March 2012, the NITE experiment proved to be a unique proposition to for NASA, since it required a trained manned operator – Jason! There was also the cost involved because the requirement that the experiment be conducted at a high latitude wouldn't come cheap. Nevertheless, the proposal won strong support within NASA which saw NITE as exactly the sort of novel research that rSLVs could enable.

Dr. Gary Thomas and Dr. Dave Fritts

Dr. Thomas has been involved in the noctilucent cloud research community since 1981 and it was he who coined the term "Polar Mesospheric Cloud," which led to units of noctilucent cloud albedo commonly described in units of "Garys" in his honor. His 1999 textbook *Radiative Transfer in the Atmosphere and Ocean* is still in use in graduate classes throughout the world. Dr. Fritts, Project PoSSUM's Chief Scientist, has guided experimental programs around the world, including rocket campaigns in Alaska, Norway, Sweden, and Brazil, radar measurements on six continents, and multi-instrument field programs.

From NITE to PoSSUM

The NITE experiment will use a camera suite comprising a group of camera systems including an HD video camera, a visible wide-field imager, and an infrared camera, all of which will be mounted in front of a trained operator. It sounds simple, but it won't be just a case of pointing cameras at clouds and pressing the shutter because there is much more that can be accomplished during these flights. To ensure they maximized their flight time, Jason's team met with researchers interested in upper-atmospheric science to see how these NASA flights could be put to best use. The idea was to create what was called an "inverse science traceability matrix": instead of starting with a list of science objectives and compiling them into a minimal number of instruments, the NITE team listed all the objectives and instruments available within the community and consolidated them in a way to best support a specific science objective that will address a major gap in our understanding of noctilucent clouds. That gap was the means by which energy and momentum are deposited and transferred into the mesosphere – a question that can be better answered through the data Lynx flights will gather and the models that can be subsequently constructed.

Gradually, Jason's project morphed into a suite of instruments that essentially converted the Lynx into an aeronomy laboratory. A science applications team was formed to address these new opportunities, and representatives were organized to address potential applications in agronomy, astronomy and astrophysics, atmospheric science, glaciology, forestry, heliosphere science, oceanography, planetary science, snow and ice science, soil science, surface tomography, and tactical planning. NITE was no longer a single experiment, but a broad, international research effort capable of broad research and extensive public outreach potential. A new name was needed, and so PoSSUM was born.

Expanding the PoSSUM Story

At the heart of each PoSSUM mission is the PoSSUMCam (Figure 9.9), an integrated camera interface tested in simulated spaceflight that supports a state-of-the-art RED video camera, the PoSSUM Wide Field Imager (WFI) camera, and a variety of automated infrared cameras. The PoSSUMCam system also supports controls that regulate other instruments, as well as the atmospheric sampler, and a user-programmable mission clock assists in situational awareness.

9.9 PoSSUMCam. Credit: Project PoSSUM

PoSSUMCam was fit-checked on the Lynx engineering model. A simple yet very operable system was needed because the scientist-astronaut operating the payload system will need to quickly identify the most germane noctilucent cloud microfeatures to track, while coordinating vehicle attitude changes with the pilot so optimal tomographic imagery can be obtained. The scientists-astronaut will have their work cut out because they will also need to trigger mission-critical events, including extending and retracting an atmospheric pressure probe and activating an atmospheric sampler when the vehicle penetrates the noctilucent cloud layer.

PoSSUM had become the first manned suborbital research program, and the human element of the program provided an opportunity to inspire audiences while communicating the science. And, as awareness of the program grew, partners became vested in the success of the program. One of these partners was ERAU – where this author works!) in Daytona Beach, Florida. In conjunction with PoSSUM, ERAU constructed a full simulation facility to support PoSSUM research, complete with a mission control center (MCC), spacesuit pressurization facility, and instrument interface. This simulation facility supported software specially developed for PoSSUM, including simulated suborbital spacecraft as well as mesospheric atmospheric models developed by Sundog Software that recreated simulated noctilucent clouds from fundamental principles of the mesosphere.

The PoSSUM Tomography Experiment

To build tomography of noctilucent cloud structures, Lynx will need to fly through the clouds, capturing as many views of the structures as possible. But these clouds are notoriously elusive and cannot be predicted. So the best chance of capturing the data will be to ensure Lynx is ready at the peak of the season and launch when strong cloud formations are observed. This will require a high steady state of readiness and a "Commit to Launch" decision will only be made when strong cloud formations are observed. And, since launch opportunities cannot be guaranteed on any specific night, the team will essentially have to "hurry up and wait."

The mission profile is unique, and was developed using calculations of atmospheric scattering on the clouds. From launch, Lynx will accelerate upwards while turning to a northern heading. The PoSSUM scientist will control the camera arrays, setting the iris and zoom of the camera systems as needed once the small structures, unobservable from the ground, come into view. A quick decision will be made on what tomographic imagery to observe and the PoSSUM scientist will need to maintain steady focus on these features while directing the pilot to make coarser adjustments in the attitude of the spacecraft, once the main engine cuts off. But the PoSSUM scientist will not only be tasked with maintaining camera systems on a specific cloud structure: as the main engine cuts off, a probe will need to be extended to record atmospheric pressures and, on penetration of the cloud layer, a sample will be taken.

After apogee, Lynx will begin to accelerate back towards the cloud layer. A second sample will be taken just before the probe is retracted prior to re-entry. Throughout all of this, a wide-field imager and infrared camera will be sequenced in close coordination with the pilot in less than four minutes of flight. The result? Lynx will bring promising new research capabilities and numerous advantages over traditional sounding rockets. For one, stabilized imagery will be possible and a well-trained operator will be able to track specific microfeatures and build tomography of these observables. Since the presence of small-scale cloud features cannot be observed from the ground and the identification of the most relevant features is somewhat subjective, a manned operator is necessary. This operator will also be able to monitor the state of health of the payloads and reduce the overall mission risk. Further, the rapid reusability of suborbital spacecraft like the Lynx will enable numerous sequential observations within a campaign season so that variations within a season may be observed and these observations can be achieved at greatly reduced cost.

PoSSUM Educational Programs

For the program to grow, more people needed to be involved. Not only would more people need to be trained to use PoSSUM instrumentation and fly PoSSUM missions, but PoSSUM presented an unprecedented opportunity to inspire and educate people about the role of the upper atmosphere because of the human involvement that aligns with the astronautics component of the program. To support this outreach, a manual entitled *The PoSSUM Scientist-Astronaut Manual* was written through contributions from team members and two programs were developed: the first, the PoSSUM Scientist-Astronaut Program, trains science and engineering professionals to communicate the program and to eventually fly PoSSUM missions. A second program, the PoSSUM Academy, serves to

9.10 Project PoSSUM 1501 Class, February 2015. Credit: Project PoSSUM

educate high-school and undergraduate-level students. Eight candidates were trained in the inaugural Scientist-Astronaut Program in February 2015: Deniz Burnham, Jamie Guined, Paul McCall, Vasco Ribeira, Jeffrey Scallon, Pete Freeland, Jonna Ocampo, and Heidi Hammerstein (Figure 9.10). Candidates trained later in October 2015 arrived from all six continents. These candidates all held a long-established desire to travel in space, but it was also evident that they wanted to contribute to actual scientific research.

The PoSSUM Scientist-Astronaut Program

The PoSSUM Scientist-Astronaut Program (see sidebar) is a five-day, fully immersive program that provides the skills required to effectively conduct research as part of Project PoSSUM. Designed and instructed by former NASA astronaut instructors and PoSSUM team scientists, this program provides high-G training, crew resource management training, spacesuit training, high-altitude training, a biometric analysis, and instruction in PoSSUMCam operations. Candidates also receive comprehensive instruction on noctilucent cloud science, observational histories, and research methods. They then learn to use real PoSSUM instruments in customized simulations of actual PoSSUM research flights to perform PoSSUM scientist-astronaut duties. These duties include the effective operation of the PoSSUMCam system, real-time identification of noctilucent cloud microfeatures of greatest scientific interest, real-time optimization of camera settings at cloud altitudes, proper use of crew resource management techniques to assure proper vehicle attitude during the mission, and the activation of instruments at cloud altitudes.

The first PoSSUM graduates had a great time following a syllabus (Table 9.1) designed to not only train them in the skills needed to become skilled PoSSUMCam operators, but also to become fully fledged suborbital astronauts. Thanks to ISS's partnership with ERAU, PoSSUM candidates were able to perform their training in high-fidelity simulators, which included the rather cozy Lynx simulator, where students operated the PoSSUMCam while wearing the latest spacesuit design courtesy of Final Frontier Design (FFD). An added bonus was the unusual attitude training (Figure 9.11) that was instructed by world champion and aerobatic legend Patty Wagstaff.

PoSSUM Scientist-Astronaut Program at a Glance

- Full tuition (includes preparatory Webinar sessions to cover academics)
- Mission simulation and crew resource management training in PoSSUMSim
- High-G flight with Patty Wagstaff using a Super Decathlon, an Extra 300
- Anti-G garment training
- High-altitude mission training in an altitude chamber
- Spacesuit training (don, doff, regulating pressure, basic mobility, fine motor skills, flight system control, contingency operations)
- Full scientist-astronaut mission simulation training in spacesuits in PoSSUMSim
- Individual instruction on PoSSUMCam and scientific video camera systems
- PoSSUM flight suit
- PoSSUM Scientist-Astronaut Manual
- Lodging and meals
- Videos of students training procedures
- Comprehensive assessment
- Graduation certificates
- Three ERAU Extended Learning credits

Curriculum

Day 1: Aeronomy, Noctilucent Cloud Science, and Aerospace Physiology Instruction
Day 2: Hypoxia Symptom Check and PoSSUMCam Operations Instruction
Day 3: Spacesuit Training and PoSSUM Mission Simulation
Day 4: High-G (Ascent and Re-entry) Operations and Biometric Systems

Prerequisites

Current FAA Class III Flight Physical
SCUBA experience
Bachelor's degree in a science engineering or technology field

Cost

US$6,000, includes all instruction, texts, and graduation

PoSSUM scientist-astronaut graduates

Class 1501: Deniz Burnham, Pete Freeland, Jamie Guined, Heidi Hammerstein, Jonna Ocampo, Paul McCall, Vasco Ribeiro, and Jeffrey Scallon.

"This is an exciting time to be involved in spaceflight, and I feel very honored and fortunate to be a part of such a special opportunity. This represents the next step in evolution and eventual routine access to space for all, and while fulfilling the dream of my lifetime, I will be able to help make a significant difference in our world through this groundbreaking research opportunity."

PoSSUM astronaut-candidate Pete Freeland, February 2015

Table 9.1 PoSSUM syllabus

Subject	*Hours*	*Description*
Introduction to Atmospheric Science		Covers the thermal structure, composition, and dynamics of the major regions of the atmosphere
Introduction to the Mesosphere		Covers observables in the mesosphere (e.g. noctilucent clouds, red sprites, blue jets, meteoric dust, airglow), physical properties of the mesosphere, density profiles, temperature, ionization, chemistry, and dynamics of the mesosphere (e.g. winds, instabilities, Gravity Waves, and Kelvin Helmholtz Instabilities)
Introduction to Global Climate		Provides a survey of global climatology, focusing on how natural and man-made changes to the atmosphere affect the global ecological system
Introduction to Noctilucent Cloud Science		Provides a conceptual understanding of the largest observables in the mesosphere and noctilucent clouds. Categorization, structure, color, polarization, formation and evolution of noctilucent clouds are covered as well as remote sensing techniques used to understand noctilucent cloud properties
Introduction to Atmospheric Scattering	1	Covers the principles governing the means by which solar radiation scatters in the atmosphere. Concepts behind Rayleigh, Mie, and geometric scattering are covered
Observing Noctilucent Clouds from Spacecraft	2	Uses principles of geometry, celestial mechanics and solar position calculations, and atmospheric scattering to provide an understanding of how PoSSUM missions are planned, including vehicle launch locations, attitude profiles, and launch-commit criteria
PoSSUM Instrumentation	2	Provides an introduction to suborbital cinematography, PoSSUMCam operations, and ground and airborne support of PoSSUM missions. Also, an introduction to space instrument design, test, and integration is provided
PoSSUM Mission Simulation	6	A survey of the principles of mission simulation and operations is provided. Students then learn to fly suborbital PoSSUM missions as the pilot and the scientist-astronaut in PoSSUMSim. Crew Resource Management (CRM) training provided by Project PoSSUM personnel: students learn how to operate PoSSUM instrumentation in a real-time analog environment
Introduction to Spaceflight Physiology	2	Provides a background on the physiological effects associated with spaceflight, including cardiovascular effects, orthoscopic hypotension, effects of acceleration, G-LOC, muscle structure and function, countermeasures, effects of spaceflight on neurovestibular system, and space motion sickness
Physiological Effects of Spaceflight	4	Suborbital missions are simulated in aerobatic aircraft to familiarize students with operations in a high-G environment. Students learn to mitigate G-induced blackouts through the use of specialized breathing techniques. Flights are conducted in a Super Decathlon aerobatic aircraft
Effects of Hypoxia		An academic background on hypoxia is provided, including an understanding of the environment, respiration and effective performance time (EPT), decompression illness, and behavioral and physiological changes as a result of hypoxia

Hypoxia Lab	Hypoxia awareness training at altitudes equivalent to 25,000 feet is provided in the PoSSUM normobaric chamber. These skills are essential to the safe operation of spacesuits
Space Life-Support Systems	Students learn the fundamentals of life-support systems, including Environmental Control and Life-Support Systems (ECLSS) and design trades associated with cabin pressure, temperature, humidity, oxygen concentration, carbon dioxide concentration, containment of hazardous vapors and particulates, and ventilation systems
Introduction to Spacesuit Operations	Students learn to don, doff, pressurize, perform safety checks, and conduct basic operations using a spacesuit
Spacesuit Operations and Mission Simulation	Students will extend upon their spacecraft and PoSSUM instrument operations skills to successfully conduct a simulated PoSSUM flight in confined environments that simulate actual missions

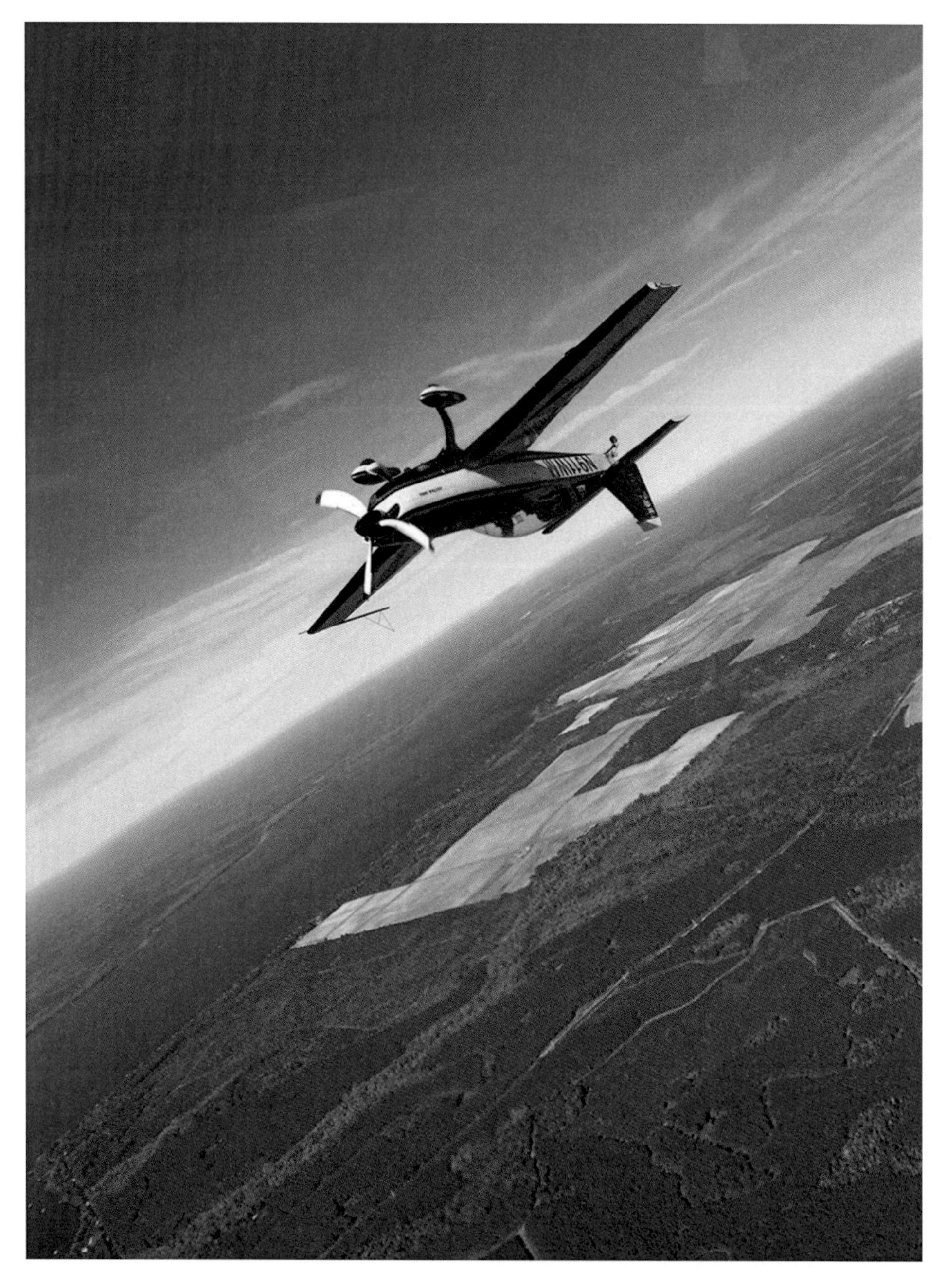

9.11 Unusual attitude training is a great way to experience a snapshot of what flying a suborbital flight will feel like. Credit: Project PoSSUM

In all of these programs, technologies and methodologies were matured as PoSSUM students also evaluated PoSSUM instruments, procedures, and spacesuits in all sorts of relevant analog environments, such as high-G flight, altitude chambers, simulated suborbital flight, and microgravity flight. Instruments had to be proven to work effectively by an operator in a pressurized spacesuit subject to high-G and microgravity environments, and several instruments had to be manipulated at the same time in these environments.

PoSSUM: A Comprehensive Suborbital Research Program

PoSSUM continually seeks to represent the science most relevant to the aeronomy community. As Lynx prepares for flight, PoSSUM is developing several instrument packages to support its research efforts. Some of these payloads have been designed for balloon applications, one of which is to launch from Alaska and the other to launch for an extended duration about the Antarctic. These missions focus on the temporal evolution of noctilucent cloud microfeatures over periods of hours or days and provide essential calibration data needed for the tomography experiments while testing and validating PoSSUM instrumentation. But, while these research efforts grow, PoSSUM's educational program continues to expand into high schools and undergraduate programs, exposing students to exactly the science and engineering that supports the research program.

Appendix I: Statement of Michael Kelly, Before the House Transportation Committee

HEARING ON COMMERCIAL SPACE TRANSPORTATION

Testimony of Michael S. Kelly,
Chairman, Reusable Launch Vehicles Working Group,
Commercial Space Transportation Advisory Committee
Before the House Aviation Subcommittee

9 February 2005

Mr. Chairman, Members of the Aviation Subcommittee, ladies and gentlemen, I appear before you today to testify on the state of the emerging private commercial space transportation industry (hereafter, "the industry"). My testimony will focus on two areas, the direction of the industry, and its plans for dealing with regulation as a result of the enactment of the Commercial Space Launch Amendments Act of 2004.

For the first time, the term "emerging" can now be used in a positive sense. It is emerging thanks to the achievements of a small group of individuals. However, make no mistake: there is as yet no industry in the sense of ongoing, revenue-producing operations. We have a long way to go before such a thing exists and even the smallest of stumbles could add years of delay.

The landmark achievement of last year was establishing a personal space flight, the conveyance of paying passengers into space, as the primary market for the industry. Any commercial enterprise requires a market, if only a perceived one, to attract the startup investment required. Because space transportation is a very capital-intensive activity, potential payoffs needed to be great. The great irony is that every visionary in this field has had, as an ultimate end goal, the establishment of a personal space flight industry. Only a tiny subset of these visionaries recognized that what was considered the "furthest out" of the markets prospects was the only sensible one with which to begin. Demand was never really in question, but it was not until Dennis Tito became the first person to pay for a trip into space that the demand was demonstrated.

E. Seedhouse, *XCOR, Developing the Next Generation Spaceplane*,
Springer Praxis Books, DOI 10.1007/978-3-319-26112-6

After that, what was required was a demonstration that private industry could develop a safe transportation system that could repeatedly take people into space, and return them safely to the earth. Last year, Burt Rutan and Scaled Composites made that demonstration, and did it so completely and decisively that even many of the visionaries in this field are stunned and amazed.

Concurrent with Rutan's demonstration, the next step in establishing an industry occurred. An operator who has the financial capacity and reputation to go the last mile stepped forth. Sir Richard Branson announced his intent to purchase several spaceships from Scaled Composites, and offer suborbital space rides to private citizens at a price of $200,000 a ticket. At last count, Virgin Galactic had 14,000 reservations. Government has kept pace with these rapid developments. The Commercial Launch Amendments Space Act of 2004, now signed into law, is the first legislation dealing with the reality of private, personal spaceflight. While the Act continued to provide for the safety of the uninvolved public, it resisted limiting the freedom of the participants in personal spaceflight. It did not attempt to legislatively preempt the right of space flyers to assess and take their own risks. It is to the everlasting credit of this Congress that these rights of the individual were explicitly acknowledge and preserved by this Act.

However, safety of space flyers is a serious issue. If it is not addressed in legislation, many asked, how would it be ensured?

Make no mistake, safety is the foremost concern of this industry. Primarily, the concern is out of basic human decency and a deep commitment to the value of human life. Close behind that motivation (though almost inseparable from it) is a more immediate concern: the economic aspect of the safety of space flyers. It is in everyone's best interest to have a safe and reliable vehicle and a safe operator in this industry.

Aviation safety has long been regulated by the federal government. But aviation safety regulations were based on the operational experience of many years. There is no such experience base for personal space transportation. A major fear of the industry, and its financial backers, has been that the government would attempt to formulate regulations in a vacuum, placing impossible obstacles in the way of people whose job is difficult enough as it is.

But the question remained of how, in lieu of government regulation, the industry would ensure the safety of space flyers. To start to find an answer, the most prominent members of the industry met in El Segundo California on 18 January of this year. Their task was to provide an industry solution to the problem of safety.

Out of this first meeting came a plan of action. The members decided to establish a federally recognized Industry Consensus Standards Organization whose purpose would be primarily to establish Consensus Standards for ensuring the safety of space flyers. If such Consensus Standards exist, they take the place of federal regulation, and provide the equivalent or greater effect.

The same will be true of spaceships and operating procedures promulgated by this future Industry Standards Organization. When faced with the choice of flying on an approved versus non-approved spaceship, a space flyer is much more likely to accept the former.

In terms of actual safety, Industry Standards are likely to be superior to government regulations. Since they come directly out of industry experience, they can be accepted and implemented quickly without the review of people who are less experienced in the field, or who have experience only in the non-applicable field of expendable launch vehicles.

Though it is too long a story to relate here, it is a matter of historical fact that aviation safety regulations have sometimes reduced safety of aircraft compared to what industry would have provided. Worse, the imprimatur of government approval carries a weight that can give a false sense of security. This violates the principle of informed consent.

Perfect safety is a worthy goal, and having it always in the forefront will keep the industry healthy and growing. However, any activity in which humans engage will someday result in an accident. There will be injuries, and there will be fatalities. No one wants this, but it will happen. How we respond is what is important. The industry has Such Standards are prevalent in the U.S. An example of how they provide safety in the face of hazards arguably greater than those posed by suborbital space flight may be found in the workings of Underwriters Laboratories. Virtually every electrical device sold in this country carries a UL stamp of approval. That stamp specifically means that committed itself to safety, and to incorporate lessons learned from such accidents as quickly and completely as possible. No one can reasonably expect more, because no more can be done.

I do not share the view of many in industry that the first fatal accident will spell the end of personal spaceflight. Such a thing has never happened in all of history, and never will happen in all of future history. It is contrary to human nature. But the outlook provided by this view is one that ensures a commitment to safety at the maximum level possible.

There is only one way to ensure perfect safety in this or any human activity, and that is to not engage in it. Legislatively, the only way to see that no one engages in an activity is to outlaw it. It is my position, and that of many in the industry and government, that federal regulation of space flyer safety would almost be the equivalent of outlawing personal spaceflight. This industry needs all of the innovation human beings can muster, and these innovations – especially those related to safety – need to be developed and implemented as quickly as humanly possible. If these things do not happen, the financial backing will be the first thing to disappear. The industry will stop "emerging," and instead submerge.

The Congress has shown its commitment to guaranteeing the freedom of this industry to grow as a commercial enterprise. That is defined as people trading value for Before I conclude, I wish to reiterate the significance of last year's events.

Without those space flights, and without the Commercial Space Launch Amendments Act of 2004, the commercial space flight industry would remain stagnant. Now it is moving forward, in a direction of which for many years we could only dream.

There are people who deserve recognition. Prior to last year, Dennis Tito proved the market for personal spaceflight by becoming the first person to purchase a ride into space. Peter Diamandis conceived and executed a brilliant plan for incentivizing the development of a private spaceship, the ANSARI X PRIZE. Paul Allen had the vision and commitment to finance such a development effort. Patricia Grace Smith, FAA Associate Administrator for Commercial Space Transportation, had the vision and commitment to help this happen while maintaining the safety of the uninvolved public. But it was Burt Rutan who, in the end, had the genius and ability to create the first private spaceship, and he who showed the world once and for all that it could be done. These are people not just of vision, but of action. They persevered in the face of obstacles that defeated others, and opened the door to the next great human adventure. I believe that humanity owes them a debt of gratitude that should and will be paid by having their names live on throughout history.

For now, we owe them – and ourselves – the commitment to work together to ensure that their accomplishments do not lay fallow. The government and industry have now defined their proper areas of responsibility. Let us preserve that, so that this great adventure may flourish.

Appendix II: President, XCOR Aerospace

Prepared Testimony to the Senate Committee on Commerce, Science & Transportation Subcommittee on Science, Technology, and Space and the House Committee on Science, Subcommittee on Space & Aeronautics Joint Hearing on Commercial Human Spaceflight

Thursday, July 24, 2003

Today I will discuss the different ways in which aircraft regulation and launch vehicle regulation protect public safety, explain why the launch vehicle approach is more appropriate for the emerging sub-orbital space flight industry, and discuss where the line between aircraft and launch vehicle regulation should be drawn. I will close with a few remarks on commercial human space flight.

A few words about my experience in this area are in order. I am President of XCOR Aerospace, an entrepreneurial space company in Mojave, California. We have been working on safe and reliable rocket propulsion systems and vehicles since 1999. I have been involved in launch vehicle regulation issues since 1998 and have been traveling to Washington regularly to work with the FAA since 2000. In the last few years, XCOR has accumulated over 1,800 firings of rocket engines without any safety issues, and we have flown a manned rocket-powered vehicle fifteen times. These early flights took place as an experimental aircraft, and we are now ready to begin construction on higher energy vehicles. We are therefore bridging the two worlds of aircraft and launch vehicle regulation.

Aircraft regulation has always developed after the fact. The first aircraft regulations did not arise until after more than 20 years and tens of thousands of flights' experience. When the first regulatory actions were taken, the operating experience of the industry was used to identify best practices and to eliminate things that didn't work.

The assumption has always been that to protect the public, we must prevent crashes. Over time, more and more such regulations have been written; usually toward a specific technology, e.g., this kind of riveting is acceptable, that kind is not. This kind of instrument is acceptable, that kind is not. After 75 years of such rule making, the aircraft industry is among the safest enterprises in the world, and also one of the most resistant to the commercial introduction of new technology. Any innovation must prove itself safer than the established practices; a difficult burden indeed, given the millions of flights' worth of

E. Seedhouse, *XCOR, Developing the Next Generation Spaceplane*, Springer Praxis Books, DOI 10.1007/978-3-319-26112-6

experience with established methods. Experimental aircraft are allowed to use new technology, but only for non-commercial applications.

Reusable launch vehicles (RLVs) are dramatically less mature. All space launches to date have been single-use expendable vehicles, except for the Space Shuttle and small sub-orbital rockets with parachute recovery. The safety record of expendable launch vehicles is poor, since a launcher with a failure rate of one in 50 is considered reliable. As a result, launch vehicle regulation has developed quite differently from aircraft regulation. In launch vehicles, we assume that failures will happen and we take steps to ensure that those failures will not endanger people on the ground. As a result, no launch vehicle accident has ever caused a casualty among the uninvolved public.

This safety is achieved by a combination of flying in sparsely populated regions and providing high-reliability means of stopping the flight if it goes awry.

In 1998, Congress expanded the regime for launch vehicles to include reusables. Since then, AST developed regulations for RLVs based on what they expected operational practices would be. It has taken four years of constant effort by AST and industry to devise and refine interpretations of those rules in the absence of precedents to point to, but we are finally getting there. Today, at least three companies, including XCOR, are going through the licensing process for sub-orbital RLVs.

The only way that the emerging RLV companies will ever be able to develop into a profitable, job-creating and tax-paying industry is to fly, and fly for revenue. And while we fly for revenue, the uninvolved public has to be kept safe. The launch vehicle regulatory regime is the only available means to protect the public while permitting revenue flight.

As recently as a year ago, I would have thought it obvious that our vehicle would be regulated as a launch vehicle. But events over the past year have shown that there are contrary opinions, which I hope we will lay to rest. The Commercial Space Launch Act of 1984, as amended, states clearly that if you have a launch license, no permission from any other executive agency is required. That language was put in place because the first attempts to launch commercially were stymied by overlapping jurisdiction; dozens of federal agencies all claimed the authority to say "no," but had no responsibility for the consequences, and hence no motive to say "yes."

Now, because some of the sub-orbital RLVs being developed have wings and pilots, some argue that these are not launch vehicles, they are airplanes. This claim is made despite the fact that NASA's Space Shuttle orbiters and Orbital Sciences's Pegasus both have wings. In 1984 Congress defined launch vehicles to include sub-orbital rockets. One might say "Well, it's a rocket, and it doesn't go to orbit, so it's a sub-orbital rocket." However, we don't want to create a loophole, in which an otherwise conventional aircraft could mount a rocket on it and claim exemption from aircraft regulation. After almost a year of work, AST proposed a new definition, in which a sub-orbital rocket is a rocket-powered vehicle whose thrust exceeds its lift for the majority of its powered flight. Since airplanes are defined as vehicles supported by lift, we think this is a good definition.

For those who have exclusively flown experimental-type aircraft, the launch vehicle regulatory world can seem daunting. On closer examination, it is less so: all that is needed is to demonstrate that the public is safe. This is only more burdensome than for experimental aircraft because the precedents are not yet set. The regulations and regime for test flying experimental aircraft are well known, and the failure modes are well explored. There are procedures for communications, emergency response, etc., written down. XCOR believes that requiring launch providers to document their procedures is worthwhile.

The largest burden in moving from aircraft to launch vehicle operation, and the least justified, is that launch providers and launch site operators have to assess their environmental impact. Aviation, including experimental aviation, operates under a categorical exclusion (CATEX) to the National Environmental Policy Act. We have discussed pursuing a CATEX with AST, but until there have been a number of reusable launch vehicles using non-toxic propellants, it is difficult to establish parameters for a category to exclude. Let me make it clear that the vehicles we and others are developing have very low environmental impact. And while the burden of documenting this is substantial, it is likely unavoidable.

Another advantage of the launch vehicle regulatory regime is that liability insurance is already established. Launch vehicles are required to carry liability insurance up to a level called the maximum probable loss (MPL). Let me make that a bit clearer. For me to launch, I have to carry sufficient insurance to cover any reasonably possible damage to third parties. The loss probability is set to a one in ten million threshold, which is so high that we could fly four times every weekday for ten thousand years before an event exceeding the MPL would occur. Only in the case of a freak accident, with losses exceeding the MPL, does the U.S. government's promise of indemnification come into play. By eliminating the need for insurance carriers to consider wildly improbable accidents in setting insurance premiums, the insurance costs to launch providers are reduced, so far at no cost to the taxpayer.

I would like to close with a few remarks on the question of carrying people in launch vehicles. Launch vehicle regulation already protects the uninvolved public. Just as with aviation in its early days, many adventurous people see this enterprise as exciting and important. They want to go. Again, just as with aviation, this enterprise will be risky and costly in its beginning; but if allowed to proceed, the cost and the risk will go down over time. We need to go through the same process as aviation; start flying, find what works and what doesn't, then make improvements. If we insist on perfect safety, we will get it because no one will ever fly.

I have been responsible for over a dozen flights of a piloted, rocket powered vehicle. I assure you that I and my engineers will fly aboard our vehicles long before we consider them safe enough for paying customers. Nor would we ever consider flying someone who was not fully informed of the risk involved. If Americans are willing to risk their lives and wealth to open a new frontier, why should we stop them.

America would not exist if our ancestors hadn't done the same. Our first flights may seem small and unimportant – but they are only the first steps on a very important road.

Appendix III: Payload Development Guide

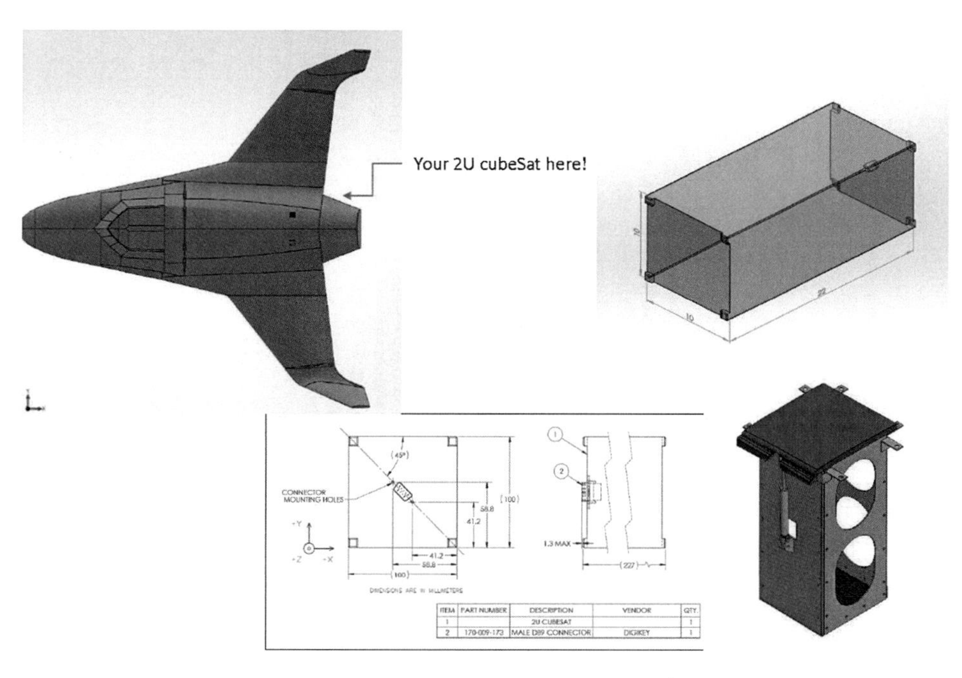

Fig. A1 Credit: XCOR

This document has been written for those interested in flying payloads on board the Lynx (Figure A1). As you know by now, each Lynx flight will be capable of carrying a number of experiments in the secondary payload carrier (SPC) behind the pilot's seat and two larger experiments in the port or starboard aft cowlings. But you can't just turn up at Midland Spaceport and request to fly a payload. There is a process to follow and that nuts and bolts of that process is outlined here.

E. Seedhouse, *XCOR, Developing the Next Generation Spaceplane*,
Springer Praxis Books, DOI 10.1007/978-3-319-26112-6

STEP 1. REGISTRATION

Regardless of whether you work for a university or intend to submit a payload as an individual, the first step you need to take is to register your intent with XCOR, the flight provider, or with Arête, Steve Heck's company (http://www.arete-stem-project.org/). At a minimum, your letter of intent should include the following:

- Description of payload and/or title of experiment
- Description of experiment
- Payload size
- Name and/or affiliation
- Name of Principal Investigator (PI)
- Name, birthdate, citizenship, and contact information for each member
- Expected payload/science development timeline
- Work plan summary

After Steve/XCOR have reviewed your letter they will get back to you to confirm whether your application to fly a payload was accepted or if supplemental information is required. This review process shouldn't take more than three or four days and can be accelerated if you provide as many reference documents (draft plans, diagrams, and procedures) as possible. Before you submit your application you may want to brush up on your knowledge of the legislation drafted by the Federal Aviation Administration's Commercial Space Transportation Office, because it is this office that has defined the eligibility requirements for payload development. You should also note that this office is particularly twitchy when it comes to experiments and payloads submitted by foreign nationals.

STEP 2. MISSION ARCHITECTURE

When submitting your payload and/or experiment, bear in mind that the Lynx is a high-performance vehicle that will subject your payload to conditions (Figure A2) much more extreme than those encountered in commercial aviation. First there is the rocket-powered take-off and ascent, which is followed in rapid succession by subsonic, transonic and supersonic speeds. Then there is the zero-g phase of the flight which is followed by an unpowered descent that will feature aerodynamic buffeting and a pullout of 4 Gs.

STEP 3. PAYLOAD VOLUME AND MASS

XCOR's smallest payload size comprises a flight-qualified 10 centimeter by 10 centimeter by 10 centimeter AMAC Plastics Model 774C clear high-density polystyrene box (Figure A3). You can fly just about anything in one these boxes as long as it doesn't weigh more than one kilogram. These boxes will fit snugly behind the pilot's seat and secured with a mounting strap. If you experiment or payload needs power the Lynx Cube structure houses a DC-DC converter that is fed by the Lynx 28V bus which steps down the power to 12 VDC.

Fig. A2 Suborbital flight will be nothing like flying commercial. Credit: XCOR

Conditions on board are shirt sleeve, with a nominal temperature of 20°C and a pressure of 72.4 kPa, although the occupants will be wearing pressure suits.

STEP 4. DESIGN CONSIDERATIONS

If you're wondering how to design your experiment or payload, one option is to download the Microgravity Development Kit (MDK) which is outlined in Appendix V. The kit, which has been developed by Terran Sciences Group (TSG – www.terransciencesgroup.com), comprises an assortment of modules that can be printed using commercial off-the-shelf 3D printers. It's a design solution that is about as low cost as they come.

4.1 Structural Integrity

Once you have the nuts and bolts of your design nailed down you will need to pay attention to the structural considerations, because your payload will be subjected to significant vibration and dynamic load on its journey to space. On this subject you should pay

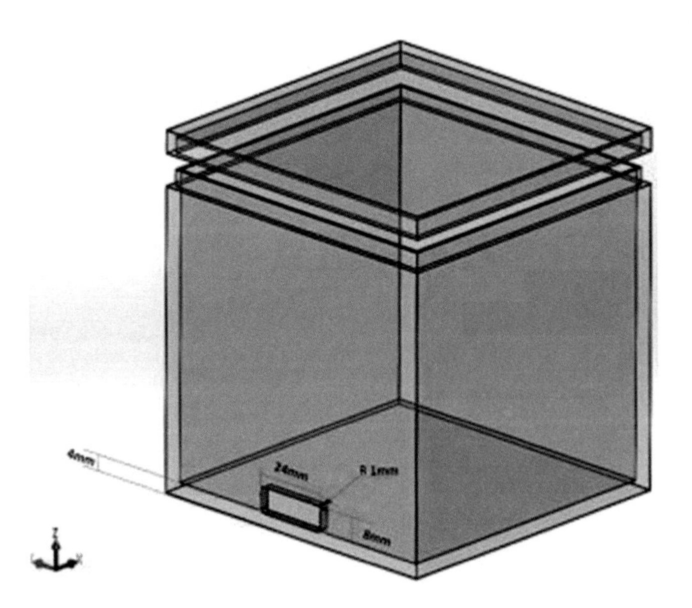

Fig. A3 A standard one unit 10 by 10 by 10 centimeter cube that can fly just about anything as long as the payload doesn't weigh more than one kilogram. Credit: XCOR

particular attention to any modifications you carry out on any modules you use because excessive drilling or cutting will likely compromise the structural integrity and result in your payload not meeting flight qualification requirements.

4.2 Containment

Built a solid module? Great! Now you can move onto the issue of containment. There are all sorts of ways to contain your payload – tethers, anti-tamper tape, seals, epoxy resin, fasteners, cables, direct mounting, encapsulation, or a combination of these methods – but no matter which method or combination of methods you use it is imperative that your payload not be compromised under acceleration, vibration or pressure ranges outside the Lynx's nominal flight envelope. To be on the safe side, you will need to design your payload to meet a factor of safety (FoS) of 2.0.

4.3 Fluids

It almost goes without saying that any payload containing fluids must – MUST – be kept in a closed tank. You will also need to ensure that and any valves and/or sensors meet the FoS of 2.0 for acceleration, vibration, and pressure differential. Chances are that if your experiment contains fluids, it will also feature pipes and/or tubes. These will likely represent a weak link in the structural integrity of your payload, so it is important that these be secured with strong structural brackets, especially in locations where connections may be subjected to asymmetric loading. If your payload also happens to have electrical components then you will need to isolate those components and may have to build in additional shielding to prevent fluid leakage.

4.4 Gases

Payload containers that contain gas – either compressed or uncompressed – will need to be hermetically sealed to prevent any escape of those gases either during nominal flight operations or during crash loading. A payload containing gas may cause you a few headaches because it must meet a FoS of at least 10.0. How do you meet this FoS? Well, you can begin by checking the US Department of Transportation (DOT) standards and apply those as a minimum. Once you're happy that you have a robust payload, it will be checked by XCOR's safety and flight personnel for final approval.

4.5 Hazardous and Radioactive Materials

Chances are that if your payload contains radioactive material or hazardous material of any kind, it will face rigorous scrutiny. I'm not saying your payload will be rejected, but it will be an uphill battle. First, what is a hazardous substance? Well, pretty much anything that is flammable, toxic or infectious. And radioactive? Well, to begin with, XCOR will not permit payloads that emit shortwave electromagnetic frequencies of 300 GHz or higher (infrared, ultraviolet, x-ray, gamma ray) beyond the confines of the payload volume. So, if you're planning on flying an imaging system of any kind, be sure that measurable leakage is prevented. The same applies to any payload that generates electromagnetic interference – this is a big red flag item because EMI interference in the VHF band could interfere with radio communication with air traffic control. If in doubt, consult your payload integration manager.

4.6 Thermal Emission, Power and Data

As you go through the process of designing your payload, bear in mind that there will be several other payloads slotted in right next to yours, some of which will contain fluids, some of which may contain gases, and many of which will contain components that generate heat. As with all the other sources of interference, the strategy you must follow to prevent your payload affecting the performance of adjacent payloads is one of isolation and shielding. Bottom line: be sure to prevent any circuitry or component from reaching a temperature of 150°C within your payload volume.

So you've shielded and isolated and pretty much made your payload as bomb-proof as possible. Now what? Well, what are the data and communications requirements for your payload? Well, first the bad news: there are no individual communications capabilities for payloads on the Lynx. The good news is that you don't have to worry about logging the flight data because this will be done for you – location, airspeed, altitude, attitude and G-loading. It will be reported for you at a 1 Hz sample rate. But what about switching your payload on and off? Well, experiments will begin their operation sequence when the payload operator triggers the "closed" or "open" condition.

Step 5. Miscellaneous Guidelines

When developing your payload, two useful reference guides that will be very helpful are the MIL-SPEC and NASA specifications for fastening, threading and mounting. If none of these methods work for securing your various components, then consider adhesives. To ensure electrical safety and to prevent loss of electrical contact under vibration and/or acceleration, locking or screw-terminal connections are probably the best way to go. And on the subject of electrical connectivity, your wiring should be bundled with ties and or wrap tubing, and all these wires should be kept out of the way of your instrumentation. Also, don't forget to factor in strain relief, especially at connectors – reckon on about three to five percent of length.

Appendix IV: The PoSSUM Suborbital Spaceflight Simulator

Vasco Ribeiro

Fig. A4 Credit XCOR

E. Seedhouse, *XCOR, Developing the Next Generation Spaceplane*,
Springer Praxis Books, DOI 10.1007/978-3-319-26112-6

This appendix explains how to perform a complete simulated PoSSUM tomography mission with Embry-Riddle Aeronautical University's (ERAU) Suborbital Spaceflight Simulator (SSFS). The SSFS (Figures A5 and A6) was designed for Project PoSSUM and simulates suborbital flight on a two-person rocketplane. The simulator runs on X-Plane software and is also used as a research tool to integrate suborbital vehicles into the National Airspace. It is located ERAU's College of Aviation and, as manager of the facility, I get the chance to fly this a few times a week. It's a lot of fun!

The simulator helps scientist astronauts practice their mission in a small commercial, two-seat, piloted space vehicle during a 30-minute suborbital flight to 328,000 ft and return safely to a landing at the runway. The SSFS has Horizontal Take off Horizontal Landing (HTHL) capability, designed for suborbital flights up to 100 km (328,000 ft) allowing it to go higher than conventional high altitude aircraft or balloons and lower than orbital spacecraft.

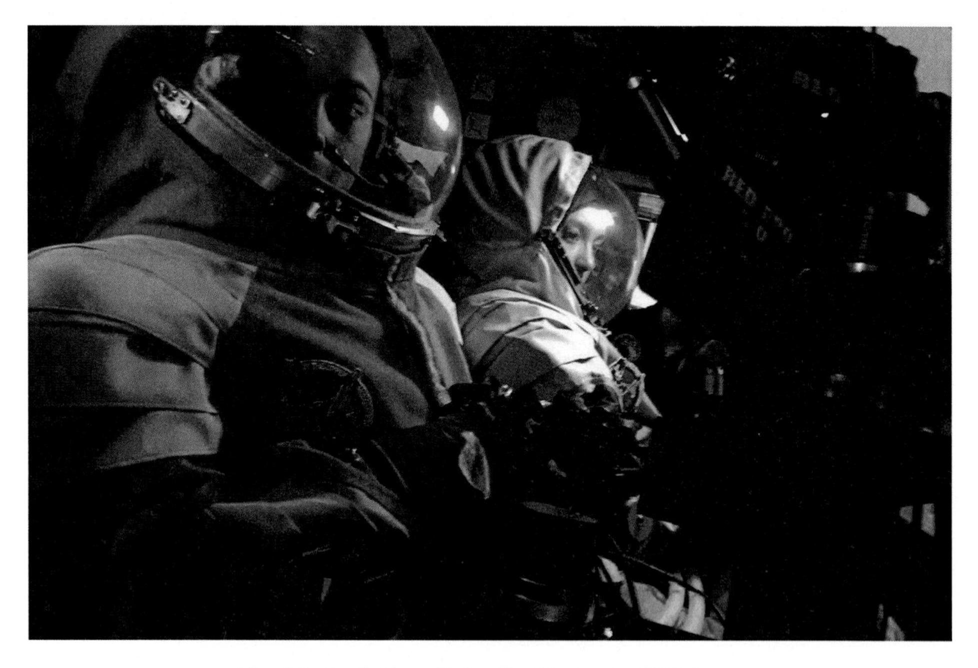

Fig. A5 Credit Jason Reimuller/Project PoSSUM

Fig. A6 Credit Reimuller/Project PoSSUM

THE SUBORBITAL SPACEFLIGHT SIMULATOR

The simulated spacecraft is based on the following assumptions:

- The wing area is sized for a gliding ratio of 8 and landing at moderate touchdown speeds near 110 knots.
- The SSFS is about 9 m (30 ft) long with a double-delta wing that spans about 7.5 m (24 ft).
- Two large fins give it good directional stability. The flight control surfaces are mixed elevator and ailerons controls (elevons) and rudder. Wing speed brakes help during re-entry to give optimal pitch.
- After landing, a drag chute is deployed to help braking.
- The empty weight is 3600 lbs and the maximum take-off weight (TOW) 10,600 lbs. Propellant load is 5650 lbs distributed across three tanks, one in the fuselage and two in the wings. Typical take-off weight is 9600 lbs, which allows it to get to 328,000 ft.
- Propulsion is provided by four rocket engines, each producing 12.9 kN (2900 lbf) vacuum thrust: this will get you to 190,000 ft at Mach 2.9 (Delta-V). After that, momentum will do the rest.

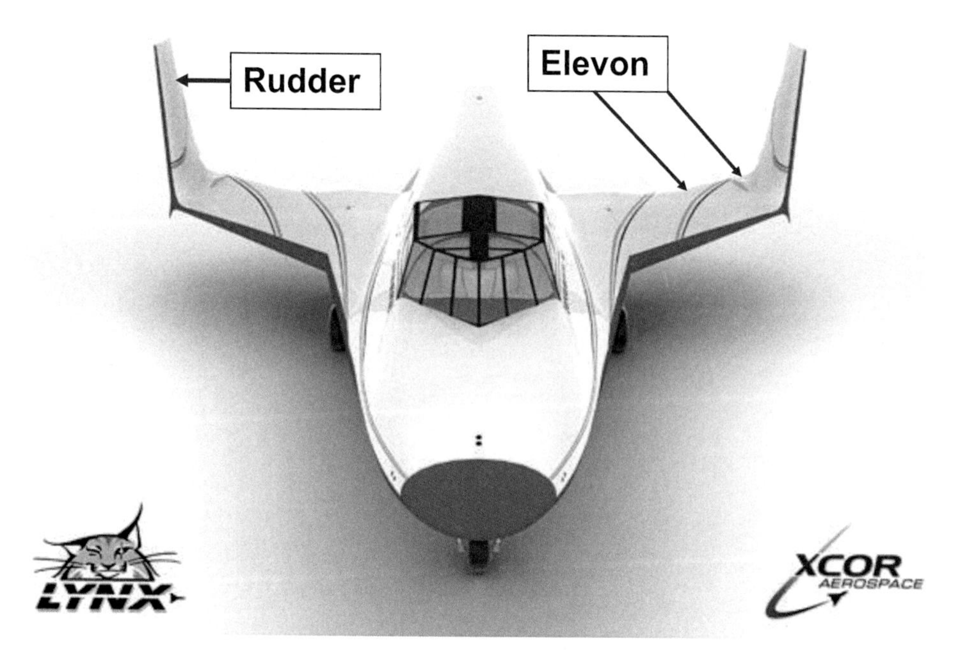

Fig. A7 Lynx control surfaces. Credit XCOR

- The model has hydraulic flight controls (Figure A7) and landing gear by means of an electric pump. In case of failure it can still be control manually. An artificial stability system helps to control the vehicle.
- A Reaction Control System (RCS) controls the vehicle in pitch, roll and yaw when in the mesosphere since flight control surfaces are inoperative at such high altitudes.

Touchscreen Panels

The PoSSUM simulator is designed to refine crew resource management techniques and skills in the manipulation of PoSSUM instrumentation in real-time in a simulated environment. The simulator has three projection screens mounted outside of a cockpit that houses two touchscreens that simulate the cockpit interior. The two bucket seats simulate spacecraft seats and are designed to secure persons wearing a Final Frontier Launch, Entry, and Ascent (LEA) spacesuit via a 5-point harness.

The two touchscreen panels (Figure A8) are present in separate screens for the pilot and co-pilot. The pilot panel has the main flight instruments and an Avidyne Primary Flight Display (PFD). The co-pilot panel has an Electronic Centralized Aircraft Monitor (ECAM), the radios (Figure A9), GPS and fuel management buttons. The PFD artificial horizon was modified to provide information on pitch angles for climbing, re-entry and gliding. A Head-Up Display (HUD) makes things easier for the pilot giving information about the flight parameters without taking the eyes from the flight path.

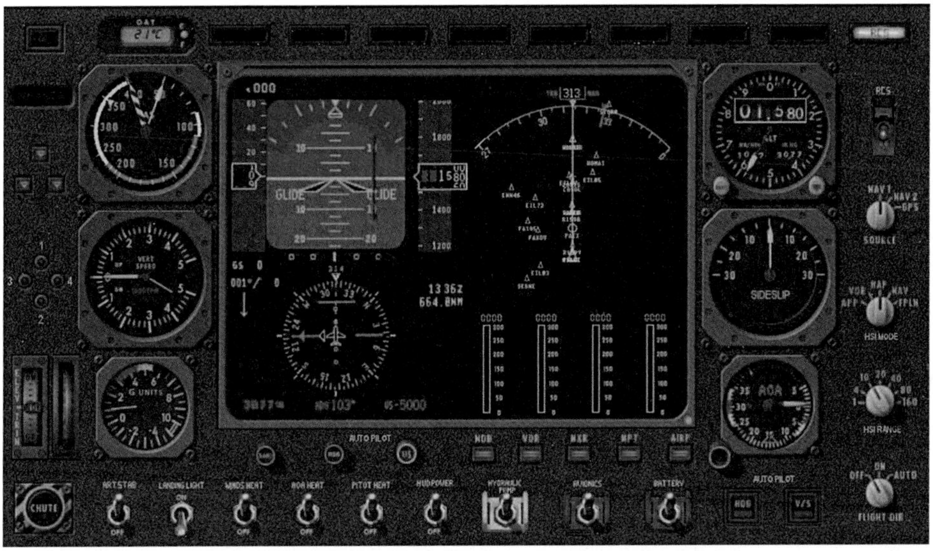

Fig. A8 Pilot panel with the central PFD. Credit Vasco Ribeiro

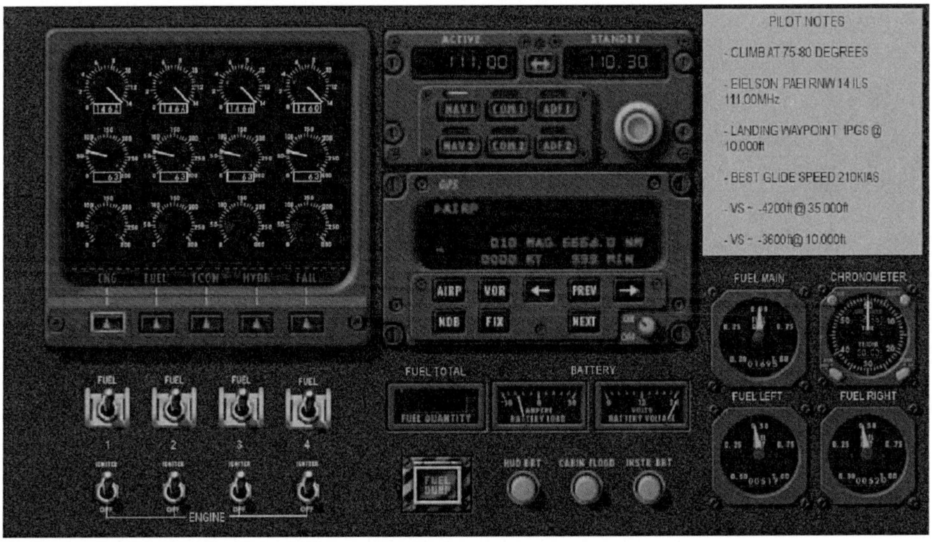

Fig. A9 Co-pilot panel with ECAM, radios and GPS, engine and fuel management buttons. Credit Vasco Ribeiro

BASELINE PoSSUM TOMOGRAPHY EXPERIMENT

This flight will depart from Eielson AFB (ICAO code PAEI) near Fairbanks, Alaska. The flight will depart Runway 32, which means that the initial heading of the spacecraft will be to the northwest. After taking off (+13 s), set the pitch attitude to degrees while making a coordinated turn to a bearing of 360° (True North). The vehicle will be under full thrust from takeoff until about 180,000 ft or more, depending on initial weight. After Main Engine Cut Off (MECO) at +183 s, the vehicle will follow a parabolic trajectory for about 3 min. Apogee will occur at about +4 m 30 s. Vehicle re-entry will occur at about +5 m 56 s, this being the most delicate part of the flight. After regaining positive control of aerodynamic flight controls, you will perform a left turn to align with waypoint IPGS (Initial Point for Glide Slope). The final approach should be at a speed of around 200 KIAS and an attitude of −8° glide slope (Figure A10). Touchdown airspeed should be at 110 KIAS (+17 m 40 s).

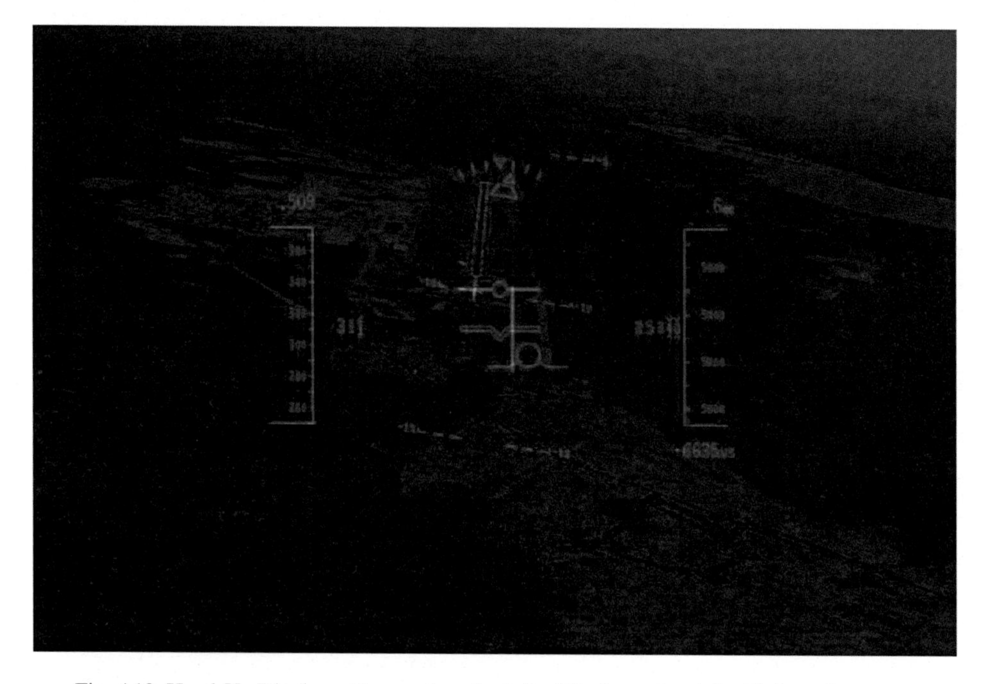

Fig. A10 Head-Up Display with speed on the left, altitude on the right, flight path vector and ILS markers on the center. Credit Vasco Ribeiro

Setting Up the Vehicle

Some settings need to be adjusted before flight:

1. Fill the tanks to 'full'. By default the fuel tanks are half-full. The payload weight bar should be set between 300 and 400 lbs. Remember that the Takeoff Weight (TOW) will determine the maximum altitude of your flight.
2. Make sure the Noctilucent Cloud Plugin is activated
3. Make sure the Date and Time are selected to 1 August, 2017 and midnight (0000 LT) respectively.

Flight Regimes

Pre-Flight Configuration for Scientist-Astronaut (MS1)

Before flight the pilot will check all systems using the check-list, setting radio frequencies for NAV and GPS waypoint navigation. Prior to the closing of the cockpit door, the scientist-astronaut will confirm all mission payloads are ready for flight. Support crew will power systems that the astronaut will not be able to manipulate once in a spacesuit.

Support Crew:	PoSSUMCam MAIN POWER—ON
	WIDE-FIELD IMAGER—Configured
	RED EPIC—Configured

Prior to launch, MS1 will engage all systems and ensure they are functional.

MS1:	WIDE FIELD IMAGER POWER—ON
	SAMPLER POWER—ON
	MCAT/MASS POWER—ON
	MISSION ELAPSED TIME—Set to Zero
	WIDE FIELD IMAGER—SEQ START
	RED EPIC—ON
	RED EPIC—REC

Lastly, just before launch, MS1 will start the chronometer:

MS1:	Mission Elapsed Time—ON

Take Off

The SSFS has three fuel tanks: a main tank in the fuselage plus two in the wings. As the wings have a lower Centre of Gravity (CG) the resultant CG will be lower, meaning that in the first stage of flight a down pitch moment will be compensated by lift of the wing. Set ¼ lower trim (nose up) for a soft take off. Give ¾ throttle. The rotation will occur at 170 KIAS. Keep in mind that the wheel gear creates resistance and a pitch down momentum.

Climbing

After take-off, increase throttle to 100 % and build speed. The climb-out will be as the vehicle is pitched to an attitude of 75–80° at an Indicated Air Speed (IAS) of around 350 KIAS. **The vehicle should never exceed 400 KIAS**. The wing fuel will burn first. After the wing fuel has burned the CG will align with the resulting vector of thrust, which will require the pilot to adjust the elevator trim. Turn the vehicle gentle to bearing 360° while on the climb. Note that the IAS will decrease as air density gets lower with altitude. Remember that your climbing pitch will give you the final base distance after re-entry. **Climbing pitch below 75 degrees will send you far away from base resulting in an emergency landing**.

Above 100,000 ft

As the atmosphere gets thinner the vehicle will accelerate to Mach 2+ and flying control surfaces will become less effective. Keep the flight path to the north and climbing pitch by means of trim. The vehicle tends to be very stable, but avoid sharp movements. At this altitude, engage the RCS system by touching the control on the monitors.

Main Engine Cut Off (MECO)

The fuel will burn to around 180,000 ft or more depending on initial weight. About 3 min will have passed since take-off. At this point the speed will be around Mach 2.9 and the vehicle will follow a parabolic trajectory, with the apogee at around 328,000 ft. To overcome the ineffectiveness of the control surfaces, the RCS helps to control the vehicle. Using these controls the vehicle may be turned in any direction using the same control. The control stick will control the RCS system just as it controls the aerodynamic surfaces at lower altitudes.

Parabolic Flight and Data Acquisition Phase

After MECO, two events need to happen:

1. The pitch-down maneuver must be initiated, and
2. The MCAT pressure probe must be deployed

PILOT:	**PITCHDOWN** to +10 degrees, Confirm Azimuth North
MS1:	DEPLOY MCAT Probe Switch—UP

The first maneuver will be a pitch down maneuver so the pitch axis is approximately +10 degrees. In this orientation the vehicle will be ascending to the North. At this time MS1 will deploy the MCAT probe and adjust the iris and zoom of the RED EPIC system.

Ascending Noctilucent Cloud Penetration

On cloud penetration the pilot will pitch down to capture the wake of the cloud. MS1 will engage the sampler to take the first of two samples.

PILOT: **PITCHDOWN** to –90 degrees,
MS1: SAMPLE Switch—ENGAGE

Apogee

At this stage of flight, which takes another 3 min, the pilot will perform experiment-specific manoeuvres without changing the flight path. The NCL experiments and camera operation will be performed at this stage. Below 250,000 ft the vehicle must be aligned with horizontal flight path (vector on Map).

Descending Noctilucent Cloud Penetration

Pitch is set to +10° in preparation of the second cloud penetration. Again, MS1 will engage the sampler to take the second of two samples.

PILOT: **PITCHDOWN** to +10 degrees,

MS1: SAMPLE Switch—ENGAGE

Re-Entry

After the second could penetration, the pilot must immediately configure for re-entry attitude.

PILOT: PITCHDOWN to –40 degrees,
Confirm Azimuth North

MS1: DEPLOY MCAT Probe Switch—DOWN

The most challenging phase of the flight occurs at Mach 3+ and very steep dive angle. While the Space Shuttle re-entry was at Mach 25 and a very flat profile while slowing down, the SSFS will fall literally to Earth, slowing down when below 120,000 ft with a 4 g deceleration and pull up for a brief period. The pitch angle should be –40° to keep good directional stability giving some 40° of Angle Of Attack (AOA). Less, and it will yaw side-to-side, back-flip (very bad for everyone on board!), and ultimately will be destroyed. Also, you will not slow down as required resulting in a higher speed and, consequently, a lower altitude before gliding which in turn will affect your gliding radius. A drag brake in the upper wing surface should be deployed prior re-entry and retrieved immediately after the vehicle pulls up. The Stall Warning Light (SWL) will pop up at 180,000 ft and then, while keeping the dive angle and slowing down, at 90,000–80,000 ft the vehicle will being the pull out. Turn off the RCS and start a left turn after the vehicle has pulled out.

Gliding

Gliding starts at around 60,000 ft when the SWL is off. Continue turning left to the waypoint IPGS bearing. At this point, calculate if you have excess altitude (energy) using the rule of thumb of one and a half mile for each 1000 ft of altitude considering the distance to base or IPGS (+8000 ft). The speed is in the range of 200–220 KIAS and −7.5° glide-slope. Use common manoeuvres to lose altitude if required, like corkscrew or S-turn. For the corkscrew keep in mind that the SSFS can lose as much as 10,000 ft in one complete turn (Figure A11). Do not slow down below 200 KIAS to lose altitude since you will sink rapidly and you may need to recover speed back again.

Landing

Get to waypoint IPGS (Figure A12) with at least 8000 ft and 220 KIAS. Use the flight path vector on the HUD to point to the start of the runway so you keep a good glide-slope. Once you have the ILS marker, use it to be sure of your relative position to the ILS glide-slope, which is −8°. This will help manage your energy. Use drag brakes if needed but keep in mind that it will pull up the vehicle. Flare before touching down and float if necessary for contact at 110 KIAS. Use drag chute and brakes to stop. Job done!

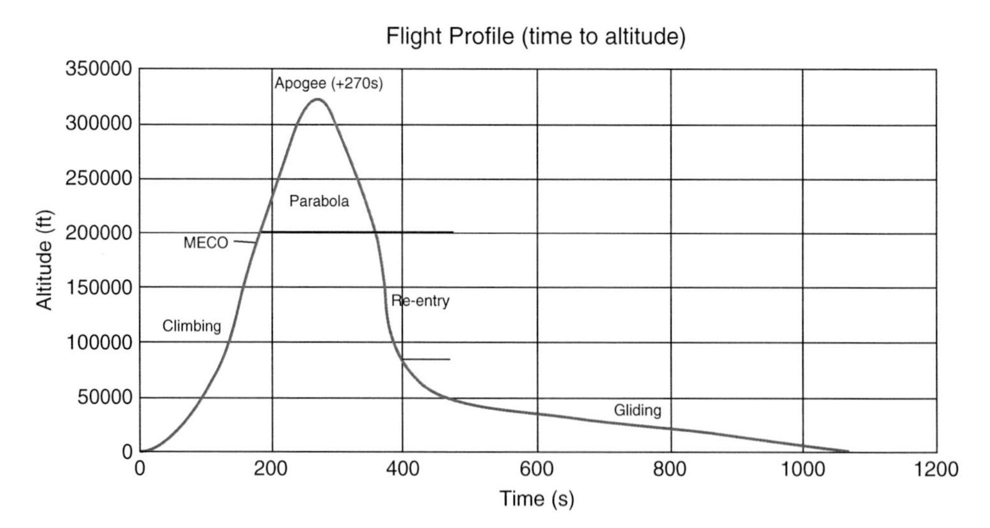

Fig. A11 Flight profile. Credit XCOR

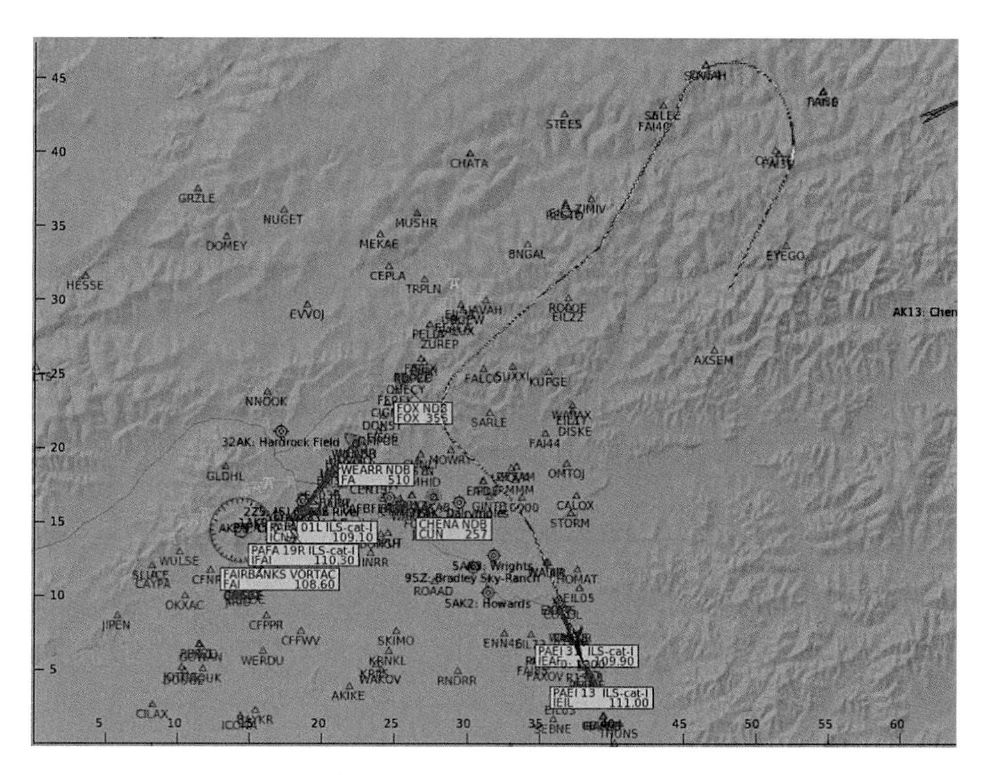

Fig. A12 Flight path after apogee. Note that the left turn was made 45 miles from base. This would require 30.000 ft altitude when starting the glide. Credit Vasco Ribeiro

Setting Up Navigational Aids (Radio and GPS)

The SSFS Simulator is equipped with radios and GPS to aid navigation even though it operates under Visual Flight Rules (VFR). The radio can hold four frequencies of each type: navigational (NAV), communications (COM) and Automatic Directional Finder (ADF). To set them first select the type then, with the selector on the right, set the frequency on Standby Display. With the outer knob, select frequency units, the inner decimal/centesimal in steps of 0.05. To make it active, just swap the position between Active/Standby. The frequencies can be obtained in the local map of X-Plane or in any airport chart.

The GPS has four types of entries: Airports, VOR's, NDB and FIX (Waypoints). The entry is an up to five-character code. Airports, AFB, and other airfields have four character codes, as PAEI for Eielson AFB or the initial point for landing (IPGS). To set just choose which type is the point on the left buttons; select the character position with the left/right arrow on the left, setting character with PREV/NEXT buttons. The bearing, distance, ground speed and Estimated Time of Arrival (ETA) to the waypoint will be displayed. Before taking off, set the radio at 111.0 Mhz (PAEI runway 14 ILS frequency) at NAV1 to

Fig. A13 Radio and GPS. Credit Vasco Ribeiro

provide custom ILS aid on approach for landing. Make sure the selector SOURCE is set to NAV1. Set GPS to waypoint IPGS so when gliding starts you can turn directly to it. You also may use the autopilot by setting heading (HDG) and vertical speed (VCS). As the vehicle operates under VFR, Eielsen AFB runway will be seen from waypoint IPGS and beyond. Again, on the pilot's PFD all the information can be displayed on the left buttons cycling through NAV, GPS and ADF (Figure A13).

Installing and Setting Up

To install SSFS files follow these steps:

1. Unzip SSS.zip to your X-Plane 10\Aircraft\Addons folder. Check X-Plane version, should be 10.30 or later.
2. For Eielson AFB scenery, unzip the file PAEI Eielsen AFB.zip to X-Plane 10\Custom Scenery folder. It uses stock objects, the add-ons are included, so you won't need to install other software.
3. Two included files have to be open to copy some lines. These will provide navigational aid. Add the lines to stock files earth_nav.dat and earth_fix.dat at XPlane 10\Resources\default data. Open with a simply text editor and copy the lines as instructed inside. Save it. WARNING! When upgrading X-Plane version these files are updated and the lines will disappear, so the process need to be re-done.
4. The earth_fix line will add a waypoint named IPGS (Initial Point for Glide Slope) situated 10 miles from Eielson aligned with runway 14. The other two lines will add a ILS direction and glide-slope with –8° at 111.00 Mhz.

Customizing Hardware Sensitiveness

To provide a better control sensation you will need to setup the joystick sensitivity to assign commands to buttons. Follow the X-Plane menus at Settings\Joystick and Equipment\Axis and Null Zone Tabs. The joystick sensitiveness is particularly important to give the proper control feeling. Read more at:

http://www.xplane.com/?article=configuring-flight-controls

Communication

Effective communication techniques must be planned and practiced between the pilot and scientist. The pilot is always responsible for the safe operation of the vehicle; failure of the pilot to accomplish this could lead to a Loss of Crew (LOC) event. The scientist is responsible to meet the science objectives of the mission; failure for the scientist to accomplish this could lead to a Loss of Mission (LOM) event. Obviously, mission safety under the pilot's judgement takes priority; however, the scientist must provide direction to the pilot as needed when changes of vehicle attitude are needed to better meet science objectives as follows:

Launch and Ascent

During launch and ascent the pilot will follow a pre-determined mission profile. In the case of the PoSSUM Tomography Experiment, the pilot will direct the vehicle to a northward heading after launch and pitch up to an 80-85 degree climb angle. The heading and climb angle are pre-determined and communication will be maintained between the pilot, Air Traffic Control, and Mission Control.

During Parabolic Flight

After MECO the scientist will be responsible for providing small attitude correction requests to the pilot, who will comply as long as mission safety is not compromised. To image microfeatures of interest small adjustments may be necessary. The pilot will always try to maintain level roll attitude, however yaw and pitch may be adjusted as needed. Remember that, under control of the RCS, roll and yaw are decoupled.

Some Comm Basics

Pitch "up" refers to 'positive pitch' and Yaw "right" refers to 'positive yaw' (Figure A14). Heading is relayed in three-digit True Headings (e.g. "zero-two-zero" for a heading of 20° east of True North). Project PoSSUM uses the following communication protocol.

PITCH UP XX DEGREES
PITCH DOWN XX DEGREES
YAW RIGHT XX DEGREES
YAW LEFT XX DEGREES

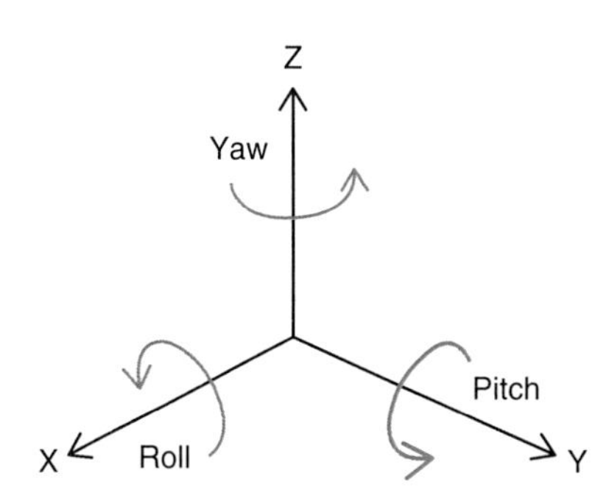

Fig. A14 Roll, pitch, and yaw. Vehicle will be ascending along the +z axis. X-axis will initially be oriented due north. Credit Vasco Ribeiro

As an example, in-cockpit communication a typical parabolic flight might proceed as:

PILOT: "MECO"
MS1: "MECO Confirmed, Pitch Down 70-degrees"
PILOT: "Roger. Pitch Down 70 degrees"

If a micro-feature of interest is right of the vehicle

MS1: "Yaw Right 20-degrees"
PILOT: "Roger. Yaw Right 20-degrees"

Ascending cloud penetration

MS1: "Cloud Penetration. Pitch Down 20-degrees"
PILOT: "Roger. Pitch Down 20-degrees"

At MECO

PILOT: "Apogee"
MS1: "Confirmed"

While descending cloud penetration

MS1: "Cloud Penetration. Pitch Up 20-degrees"
PILOT: "Roger. Pitch Up 20-degrees"

At minimum safe altitude to configure for re-entry

PILOT: "Re-entry Attitude"
MS1: "Roger."

Re-entry and Landing

The pilot will configure the vehicle to re-entry attitude once the vehicle descends to the necessary altitude, at which point no further attitude requests will be considered. Again, the re-entry angle and glide approach path are pre-determined and communication will be maintained between the pilot, Air Traffic Control, and Mission Control.

PoSSUM PILOT-ASTRONAUT MISSION CHECKLIST

BEFORE FLIGHT
Battery.................................ON
Avionics...............................ON
HUD.....................................ON
Landing lights......................OFF
Windshield heat.................ON
AOA sensor heat.................ON
Pitot tube heat....................ON
Hydraulic pump...................ON
HUD bright.....................AS REQ
Instruments bright........AS REQ
Cabin flood....................AS REQ
RCS.....................................OFF
Art stab................................ON
Trim..................................CHECK
Brakes..................................ON
Brake Chute........................OFF
CG display.......................CHECK
Altimeter...CHECK BARO PRESS
Selector SOURCE NAV1..CHECK
GPS......................SET WPT IPGS
Radio........................SET NAV 1
Radio.......STANDBY 111.00Mhz
Radio...........ACTIVE 111.00Mhz
Avidyne PFD..............SET NAV 1
Avidyne PFD.................SET GPS

ENGINES START
Fuel Tank selector................ALL
Throttle lever.......................IDLE
Turn fuel engine 1................ON
Turn fuel engine 2................ON
Turn fuel engine 3................ON
Turn fuel engine 4................ON
Start engine 1.............ENGAGE
Start engine 2.............ENGAGE
Start engine 4.............ENGAGE
Start engine 4.............ENGAGE
Engine lights.........................ON
MFD systems.................CHECK
Elevator trim.................1/4 UP

TAKE OFF
Brakes.................................OFF
Throttle..............................100%
Rotate.....................170 KNOTS
Landing Gear........................UP

CLIMBING
Pitch 75+ degrees...CHECK TRIM
Bearing 360 degrees........CHECK
MECO at 190,000 ft...........CHECK
Speed Mach 3..................CHECK
RCS...ON

PARABOLA LESS 250,000 ft
Map dir. vector forward..CHECK
Drag Brakes....................DEPLOY
Elevator trim....................1/4 UP
Descent Pitch -40.............CHECK

RE-ENTRY LESS 200,000 ft
Stall Warning Light ON.....CHECK
Descent Pitch -40..............CHECK
AOA 40 degrees...............CHECK
Speed Mach 3+.................CHECK
G accelerometer 4g..........CHECK

GLIDING
Stall Warning Light OFF....CHECK
RCS..OFF
Drag Brakes...................RETRACT
Elevator trim....................AS REQ
Speed below 45,000 ft..220 KIAS
Glide Slope 7-8 degrees....CHECK
Waypoint bearing.............CHECK
Altitude...............................CHECK
Waypoint distance............CHECK
Glide Slope radius.............CHECK

WAYPOINT IPGS
Altitude > 8000ft...............CHECK
Speed > 200KIAS...............CHECK
ILS markers.........................CHECK

FINALS
ILS markers.........................CHECK
Radio altimeter (HUD)......CHECK
Drag brake........................AS REQ
Landing gear.....................DOWN
Elevator trim.....................AS REQ

LANDING
Speed 160KIAS..................CHECK
Flare..................................CHECK
Touch down 110KIAS........CHECK
Drag chute...............................ON
Brakes.....................................ON

PoSSUM SCIENTIST-ASTRONAUT MISSION CHECKLIST

Prior to Cockpit Door Closing:

PoSSUMCam MAIN POWER	ON
WIDE-FIELD IMAGER	CONFIGURED
RED EPIC	CONFIGURED

Prior to Launch Commit Decision:

MCAT + MASS POWER	ON
SAMPLER POWER	ON
WIDE FIELD IMAGER POWER	ON
RED EPIC	ON
Mission Elapsed Time	SET TO ZERO

Prior to Takeoff:

WIDE FIELD IMAGER	SEQ START
RED EPIC	REC
Mission Elapsed Time	ON

Take off

RED EPIC	ADJUST IRIS AND ZOOM

Climbing

RED EPIC	ADJUST IRIS AND ZOOM

Main Engine Cut Off (MECO)

MCAT Probe Switch	EXTEND
RED EPIC	ADJUST IRIS AND ZOOM

Ascending Noctilucent Cloud Penetration:

SAMPLE Switch	ENGAGE
RED EPIC	ADJUST IRIS AND ZOOM

Apogee:

Mission Elapsed Time	NOTE
RED EPIC	ADJUST IRIS AND ZOOM

Descending Noctilucent Cloud Penetration:

SAMPLE Switch	ENGAGE
RED EPIC	ADJUST IRIS AND ZOOM

Re-entry

MCAT Probe Switch	RETRACT

After Landing:

Mission Elapsed Time	NOTE
MCAT + MASS POWER	OFF
SAMPLER POWER	OFF
WIDE FIELD IMAGER	OFF
RED EPIC	OFF

Appendix V: Microgravity Experiment Developer's Kit

Justin Karl

Microgravity Experiment Developer's Kit

What is it?

A collection of components, parts, and microcontrollers that make designing and building a flight-qualified microgravity experiment easier for everyone. With Citizens in Space, YOU CAN FLY A PAYLOAD ON A REAL SPACE FLIGHT FOR FREE!

Who can use it? Is it expensive?

Anyone can use it! Not to mention, participants in the Citizens in Space project can simply download and make 3D-printed parts for free. If you need help, the kit's provider will do so at cost.

About the project:
http://www.citizensinspace.org

Kit, developer guides, and help:
http://www.terransciencesgroup.com/CISP

Submit your experiment idea:
experiments@rocketacademy.org

E. Seedhouse, *XCOR, Developing the Next Generation Spaceplane*,
Springer Praxis Books, DOI 10.1007/978-3-319-26112-6

Microgravity Experiment Developer's Kit

What kind of experiments can I make?

All sorts! Anything you can fit it inside the 4"x4"x4" box (or larger if your experiment merits a 2U or 3U space) that isn't dangerous or exceeds mass requirements can fly. The pieces in the graphic represent a biological experiment where specimens are observed with a consumer-grade USB microscope connected to a TI BeagleboneBlack. These parts only cost about $200 total!

Part types:

- Power distribution
- Microcontrollers
- Cameras and microscopes
- Specimen containers
- Optics (mirrors, lenses)
- Light sources
- Fluid containment
- Structural

More added all the time... you can request something new, too!

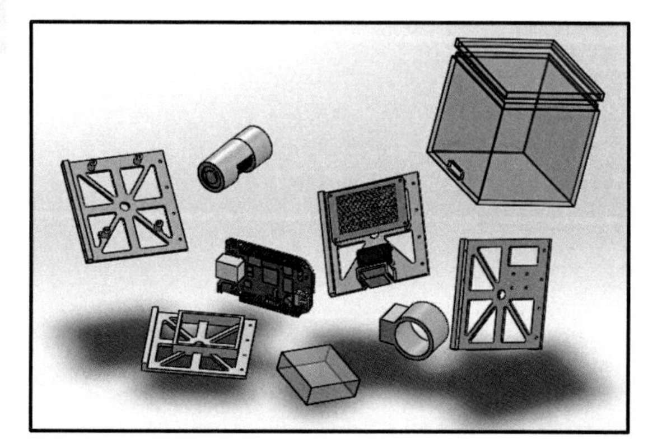

Microgravity Experiment Developer's Kit

How do I use the kit?

It's easy... Each module is a panel, and four of them fasten together with screws to make a structural assembly that fits into the primary payload container box. Choose modules that support what you need for your experiment and put them together- that's it.

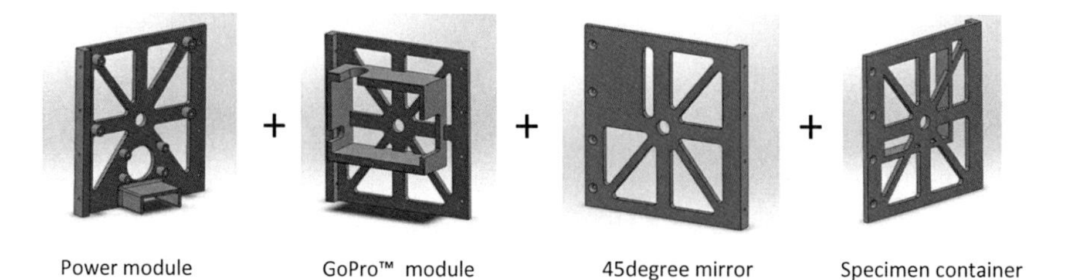

Power module | GoPro™ module | 45degree mirror | Specimen container

= your microgravity video observation experiment!

Microgravity Experiment Developer's Kit

What's available on the kit site?

On the site you can download STL or SolidworksPRT files, as well as PDF's of the datasheets that accompany most modules.

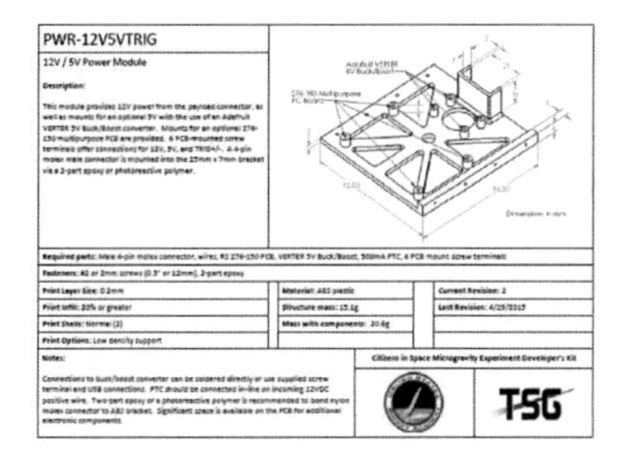

Datasheets have helpful information such as mounting point locations for popular components, data about which modules do and don't fit together, and location options for components that could be situated on different parts of the module.

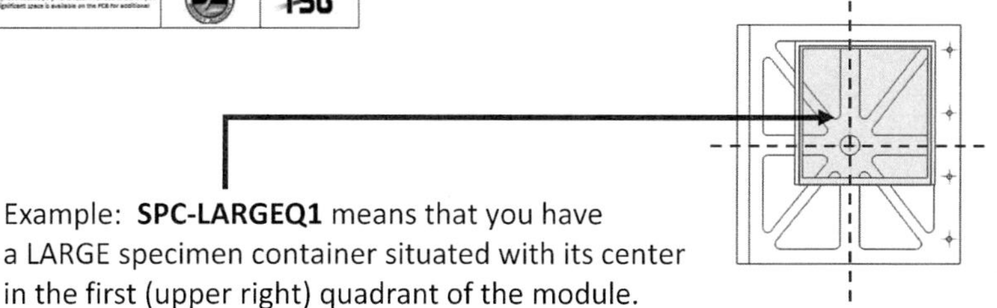

Example: **SPC-LARGEQ1** means that you have a LARGE specimen container situated with its center in the first (upper right) quadrant of the module.

Microgravity Experiment Developer's Kit

I want to get a flight slot! How do I get started?

1. Think of a good experiment! It doesn't have to be complicated to get useful results.

Starter ideas:

- Melt and solidify chocolate in microgravity
- Electrolysis in zero-g
- Epoxy curing in zero-g
- Outgassing monitoring
- Soldering in zero-g
- Cabin temperature/pressure measurement
- Photoreactive resin in zero-g
- Seed/plant biology and genetic effects
- Animal behavior in zero-g (Tardigrade or smaller, please!)
- Hardware qualification for longer flights
- Liquid mixing
- Dust/particle impacts
- Dust/particle settling

Microgravity Experiment Developer's Kit

I want to get a flight slot! How do I get started?

2. Look through the **Payload Design and Manufacturing** guide (PDM) at http://www.terransciencesgroup.com/CISP

   This guide will include information about the flight itself as well as **what information to put in your Letter of Intent.**

3. **Send that letter!** A letter of intent to experiments@rocketacademy.org gets the formal process started.

4. **Start building!** When you get a confirmation email, you can work at your own pace. The Payload Officer (PO) will help you through the process if you need it, but the important thing is to follow the guides and stay in communication.

Microgravity Experiment Developer's Kit

- Hundreds of standardized components for use with US Rocket Academy's payload volumes and carrier rack
- Uses 3D-printed structural components designed to be printable by consumer-grade and DIY 3D printer builds
- Supports use of inexpensive, popular components and hardware that are readily available
- Printing/assembly/advice available at cost for groups without access to resources
- Licensing is free for citizen scientists

Part files, STL files, documentation, and payload planning information available:

terransciencesgroup.com/CISP

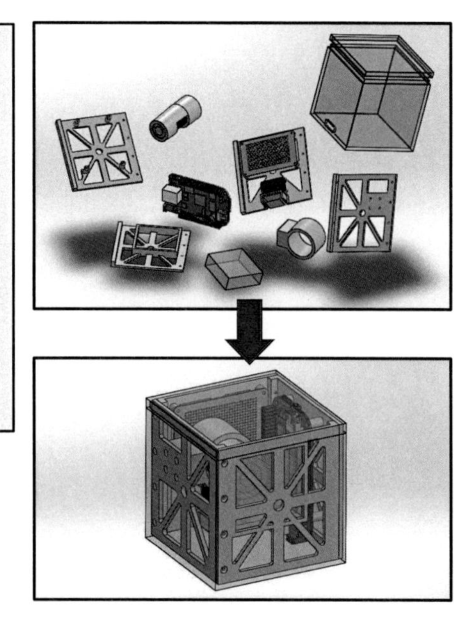

Professionals, hackers, K-12, universities...
send your intent to participate today! **experiments@rocketacademy.org**

Appendix VI: Neurological Assessment for Suborbital Crewmembers

Title	Neurological Assessment for Suborbital Crewmembers
Sponsor	Embry-Riddle University
Identifier	Neurological Function
Category	Medical
References	• Suborbital Astronaut Evaluation Document • Preflight, Inflight and Postflight Medical Evaluation for Suborbital Flights
Purpose	Perform a neurological assessment that provides a test of neurosensory adaptation to Earth gravity following suborbital flight
Measurement parameters	Balance control. Sensory-motor integration
Deliverables	Test results can be used by flight surgeons to determine performance decrements following suborbital flight
Flight duration	20 minutes
Flight characteristics	Preflight, inflight and postflight data collection

E. Seedhouse, *XCOR, Developing the Next Generation Spaceplane*,
Springer Praxis Books, DOI 10.1007/978-3-319-26112-6

Preflight training description	A familiarization session to be conducted 20 days before launch to negate learning effects. An astronaut who has previously performed the protocol will not be required to perform it again. At the test facility the astronaut will change into shorts and socks and will report pharmaceutical and alcohol consumption, physical activities, eating and sleeping schedules within the previous 24 hours. Baseline foot and hip measurements will be performed. The astronaut will be instrumented with motion markers on the lower legs and hips, and will wear a safety harness and headphones. The astronaut will complete a clinical battery of sensory organization tests (SOTs) provided by the Equitest system. Three randomized trials of six sensory organization tests will be performed (upright posture with normal, absent, and/or mechanically altered visual and proprioceptive sensory inputs). Each test lasts for 20 seconds during which the astronaut must maintain normal stance. Measured performance will be based on peak body sway during each test. The astronaut will perform 3 trials each of 4 static head tilt and dynamic head movement tests, each lasting 20 seconds. Finally, the astronaut will perform 4 platform translation and 5 platform rotation tests, each lasting 10 seconds. When testing is complete the astronaut will be de-instrumented.
Schedule	The clinical SOT test conditions are (3 trials each): 1. Standard eyes open Romberg test-visual surround and support surface fixed 2. Standard eyes closed Romberg test-support surface fixed 3. Sway-referenced visual surround with fixed support surface 4. Sway-referenced support surface with fixed visual surround 5. Sway-referenced support surface with eyes closed 6. Sway-referenced support surface with sway-referenced visual surround The head movement test conditions are (3 trials each): 1. SOT 2 (above) with static head tilt back (eyes closed) 2. SOT 5 (above) with static head tilt back (eyes closed) 3. SOT 2 (above) with dynamic head pitch (backward and forward, eyes closed) 4. SOT 5 (above) with dynamic head pitch (backward and forward, eyes closed) The MCT test conditions are: 1. Eyes open, small amplitude forward translation (2 trials) 2. Eyes open, large amplitude forward translation (2 trials) 3. Eyes open, toes-up support surface rotation (5 trials)

Ground Support Requirements	Preflight Hardware	Preflight Software	Test Location	
	Equitest Posture Platform, Optotrak Motion Analysis System, Equipment Racks, Subject Safety Restraint System, Test Supplies, Tone Generation Equipment, Universal Power Supply	N/A	Simon Fraser University	
Training Facilities	**Room Dimension**	**Electrical Outlets**	**Temperature**	**Lighting**
	3 × 5 × 3 meters	2 Electric Outlets with Amp Rating of 120V, 20A, 60Hz. Note: Each circuit should be accessible through a standard three-wire, grounded, duplex receptacle located within 2 meters of the platform.	20–22°C	Standard
	Hot/Cold Water	**Privacy**	**Other**	
	No	2 Test operators	2 chairs. 1 table	
Constraints	Limit participation in provocative training exercise (i.e., flight motion simulators, centrifuge simulations, training aircraft) for 24 hours prior to testing; limit alcohol consumption for 24 hours prior to testing. No maximal exercise 4 hours prior to test. No drugs that affect sensorimotor performance.			
Notes	Test Termination Criteria: 1. Syncope or significant pre-syncopal symptoms 2. Vomiting or significant motion sickness symptoms 3. Significant foot tenderness or muscle soreness 4. Subject requests to stop			

Postflight activities	The astronaut will change into shorts and socks. The astronaut will be instrumented with motion markers on the lower legs and hips, and will wear a safety harness and a pair of headphones. The astronaut will then complete a graded and reduced set of tests, detailed below. If the astronaut is unable to complete or falls during a sub-set of tests he/she will move on to the next sub-set. If the astronaut falls twice, the test session will be terminated (usual test termination criteria also applies).
Schedule	The clinical SOT test conditions are (2 trials each): 1. SOT 1 2. SOT 3 3. SOT 4 4. SOT 2 5. SOT 5 The HM test conditions are (2 trials each): 1. SOT 2 with static head tilt back (eyes closed) 2. SOT 5 with static head tilt back (eyes closed) 3. SOT 2 with dynamic head pitch (backward and forward, eyes closed) 4. SOT 5 with dynamic head pitch (backward and forward, eyes closed) The MCT test conditions are: 1. Eyes open, small amplitude forward translation (2 trials) 2. Eyes open, large amplitude forward translation (2 trials) 3. Eyes open, toes-up support surface rotation (5 trials)

Duration	Schedule	Flexibility	Personnel
20 minutes	L – 10 days		2 test personnel

Ground support requirements	Preflight Hardware	Preflight Software	Test Location
	Equitest Posture Platform, Optotrak Motion Analysis System, Equipment Racks, Subject Safety Restraint System, Test Supplies, Tone Generation Equipment, Universal Power Supply	N/A	Simon Fraser University

Training facilities	**Room Dimension**	**Electrical Outlets**	**Temperature**	**Lighting**
	3 × 5 × 3 meters	2 Electric Outlets with Amp Rating of 120V, 20A, 60Hz. Note: Each circuit should be accessible through a standard three-wire, grounded, duplex receptacle located within 2 meters of the platform.	20–22°C	Standard
	Hot/Cold Water	**Privacy**	**Other**	
	No	2 Test operators	2 chairs. 1 table	
Constraints	Limit participation in provocative training exercise (i.e., flight motion simulators, centrifuge simulations, training aircraft) for 24 hours prior to testing; limit alcohol consumption for 24 hours prior to testing. No maximal exercise 4 hours prior to test. No drugs that affect sensorimotor performance.			
Notes	Test Termination Criteria: 1. Syncope or significant pre-syncopal symptoms 2. Vomiting or significant motion sickness symptoms 3. Significant foot tenderness or muscle soreness 4. Subject requests to stop			
Data delivery	Test results will be delivered to the flight surgeon via oral report by designated project lead.			

Index

E. Seedhouse, *XCOR, Developing the Next Generation Spaceplane*, Springer Praxis Books, DOI 10.1007/978-3-319-26112-6

Printed in the United States
By Bookmasters